NRA
National Rivers Authority

CONTAMINATED LAND AND THE WATER ENVIRONMENT

Report of the
National Rivers Authority

March 1994

Water Quality Series No. 15

LONDON: HMSO

PREFACE

There is nothing new about contaminated land: what this report does provide is a new look at it - specifically from the viewpoint of the extent to which it does, or is likely to, result in the pollution of surface and groundwaters of England and Wales. The NRA has already examined systematically the principal causes of poor water quality: the manner by which discharges from point sources are consented, the extent and origin of accidental inputs, and the complex relationships between agriculture and water quality.

Consideration has also been given to objectives which could be set by statute to maintain, and where necessary improve, water quality, and to the steps which need to be taken generally to protect groundwater from pollution. An evaluation of the total quantities of a number of substances entering coastal waters from land-based sources is also being undertaken. None of the aims of these initiatives will be fully met unless the role played by contaminated land is evaluated, and the means by which the problem can be addressed. This, however, is a complicated task because the NRA is not the only organisation involved; indeed it is not even the lead organisation. This report therefore explains these issues and considers how best to proceed. Emphasis is placed on two separate aspects: how best to ameliorate the problems arising from the past, and how best to prevent similar occurrences in the future.

CONTENTS

		Page
	EXECUTIVE SUMMARY	1
1.	**INTRODUCTION**	3
2.	**WHAT IS CONTAMINATED LAND?**	4
	NRA Interest	4
	Contaminated Land - A brief history	5
3.	**CONTAMINATED LAND AS A PROBLEM - THE UK EXPERIENCE TO DATE**	7
	Experience Abroad	8
	Contaminated Land in the Netherlands	8
	The USA and Superfund	9
4.	**WATER POLLUTION - ACTUAL AND POTENTIAL**	10
	General Risks	10
	General Risks - Re-development of Contaminated land sites	10
	Examples of Actual pollution	11
	Landfill Sites	11
	General Industry	12
	Agriculture and Horticulture	15
	Mine Spoil Tips	15
5.	**THE LAW**	17
	Statutory Duties and Powers of the NRA	17
	Direct Powers - (i) Ameliorating Water Pollution	18
	Direct Powers - (ii) Prevention of Future Problems	19

	Discussion	19
	Indirect role of the NRA	20
	The Environmental Protection Act 1990	20
	Section 33 and 36: Treatment or Disposal of Waste, and Licensing	20
	Section 34: Duty of Care	20
	Section 61: Clean-Up Liability	21
	Section 143: Registers of Land which has been put to a Contaminative Use	21
	The Planning System	23
	The Planning System and the NRA	23
	The Planning System and Contaminated Land - How it works	24
	Recent guidance on the Planning System	29
6.	**PAYING FOR CLEAN-UP**	**30**
	Derelict Land Grant (DLG)	30
	Operational Priorities and Objectives	31
	How DLG currently works	32
	City Grant	33
	Urban Regeneration Agency	33
	Supplementary Credit Approvals (SCAs)	33
7.	**THE NRA - NATIONAL AND INTERNATIONAL RESPONSIBILITIES**	**34**
	International Responsibilities	34
	Groundwater	35
	Draft Landfill Directive - COM(91) Final	37
	Draft Directive on Civil Liability in Waste -COM(91)219	37
	Surface Waters	38
	Dangerous Substances Directive (76/464/EEC)	38
	PARCOM and the North Seas Interministerial Conferences	39

8. THE JOB TO BE DONE — 40

Introduction — 40

Cleaning up the Past — 40

(a) Estimating the Scale of the Problem — 40

(b) Estimating the Nature of the Problem — 41

(c) Dealing with the Problem — 42

Knowing what best to do - R&D — 43

Prevention in the Future — 44

Statutory Consultations — 44

Regulations — 44

Groundwater Protection Policy — 45

Conclusions and Summary — 46

Assessment — 46

Remedial Action — 46

Other Actions — 47

APPENDICES

Appendix One — Water Resources Act 1991 — 49

Appendix Two — Proposals for EPA'90 Section 143 Registers — 55

Appendix Three — Current and Proposed NRA R&D Projects — 57

REFERENCES — 59

LIST OF TABLES

Table 1:	Uses of land which may give rise to contamination	6
Table 2:	Examples of Clean-Up Policy in Four Countries	9
Table 3:	Water quality of Brightley Stream during heavy rainfall, October 1989 (from NRA WQ Series No 6 1992)	15
Table 4:	Management Steps and Associated Legal Powers	17
Table 5:	NRA Consultee Status in the Planning System	23
Table 6:	DLG - How it Works	32
Table 7:	Current and draft EC Directives related to Water Pollution from Contaminated Land	35
Table 8:	List I and List II - Dangerous Substances	36

LIST OF FIGURES (COLOUR PLATES)

Figure 1:	The above ground view of the old landfill site	25
Figure 2:	Removal of cyanide during remedial works	25
Figure 3:	Above ground view of the old waste tip site	25
Figure 4:	Discharge point to local watercourse now contaminated with PCBs	25
Figure 5:	Redeveloped landfill site - now an industrial estate	26
Figure 6:	Unvegetated land on derelict extraction and refining site for heavy metals	26
Figure 7:	Surface run-off giving rise to gross pollution potential in the River Tamar	26
Figure 8:	Refinery site and adjacent forest	26
Figure 9:	Blyth Harbour, where high levels of PCBs have caused extensive pollution	27
Figure 10:	Containing the pollution at Blyth Harbour	27
Figure 11:	Waste briquettes from the zinc smelting process - used to raise banks on part of the site	27
Figure 12:	Surface water discharges - en route to Severn Estuary	27
Figure 13	"Main Drain" - seepage into River Tame	28
Figure 14	Seepage of contaminated water into River Tame	28
Figure 15:	Daffodil growing areas, now contaminated with biocides	28

EXECUTIVE SUMMARY

Although the subject of contaminated land evokes a range of responses and reactions, it is by no means clear as to what contaminated land actually is, and how it relates to, for example, derelict land. This report therefore first explores the subject in general and how it has been approached both in the UK and abroad.

The NRA's interests are directed towards the extent to which contaminated land already does, or has the potential to, cause pollution of controlled waters. In general, the risk of pollution arises from a very large range of contaminative uses to which land has been put, and increases particularly when contaminated sites are to be re-developed. Examples of sites causing actual pollution around the country are examined in some detail in order to explore the nature and extent of the problem. They arise from completed landfill sites, waste tips and other such areas. They arise from past industrial sites used for engineering works, manufacturing, and the chemical industry - even ship breaking. They also arise from agriculture and horticulture.

The NRA does of course have duties and powers with respect to pollution of the aquatic environment, and the extent to which these can be applied either directly through the Water Acts of 1989/1991, indirectly through the Environmental Protection Act of 1990, or through the Acts relating to the planning system, have to be fully considered. There is also the question of who is responsible, and who pays to clean up such land when it is deemed necessary to do so. The use of the Derelict Land Grant, City Grant, and Supplementary Credit Approvals all have to be considered alongside the NRA's own limited ability to contribute. And there are national and international obligations to consider, which may dictate priorities with respect to how the money should best be spent. These not only include Directives from the European Commission, but commitments arising from the Paris Commission, and from Inter-ministerial Conferences on the North Sea.

So what needs to be done? With regard to cleaning up the past, the NRA has yet to evaluate fully the scale of the problems, and the nature of the problem. Nevertheless it is suggested that, via Catchment Management Plans, contaminated areas which are currently causing breaches of water quality standards, or contribute to a significant (> 1% of the total annual) input of persistent and toxic substances into coastal waters, or are a source of more than trace quantities of PCBs, be identified now. Similarly, an evaluation of such contaminated areas known to be, or at risk of, contaminating groundwater needs to be undertaken in conjunction with the NRA's Groundwater Protection Policy.

There are also certain steps which could be taken towards dealing with the problem. The first approach should be to those responsible for the pollution, and the second to bring pressure to bear on others who have a primary responsibility, and the money, to deal with it. There are also more direct steps which the NRA could take; these include the expenditure of money on capital only projects, providing that they do not require a commitment to financial support in order to run them. Such money, would, in any case, require prior approval of the Department of the Environment when in excess of £0.5m. And if public money was to be spent in such a manner, it should be justified on the basis that it was in wider interests of the NRA's water resource management responsibilites to do so.

Prevention is naturally always better than cure, and the NRA thus needs to consider further how best it can use its direct responsibilities with regard to water quality, its indirect powers, the scope for further regulations - if necessary - and the full implementation of its Groundwater Protection Policy.

1. INTRODUCTION

1.1 Deterioration of water quality can result from a number of activities; the NRA has been addressing these in a systematic way. Many contaminants arise from point source discharges, controlled by the consents system. Many arise from diffuse sources, and the agricultural proportion of these was addressed in the report "The Influence of Agriculture on the Quality of Natural Waters in England and Wales" (NRA, 1992). But this still leaves other sources of diffuse and accidental discharges to be examined. Some of these actual and potential discharges come under the category of contaminated land, and thus the NRA cannot plan to improve water quality without addressing this issue. The setting and implementing of Water Quality Objectives (WQOs) on a statutory basis for controlled waters in England and Wales over the next few years, within a catchment management framework, will also accelerate both the growing awareness of the threat which contaminated land can pose to the water environment, and the increasingly urgent need to manage the problem.

1.2 Over the last few years the subject of contaminated land has become one of the most debated environmental issues in the UK. As a result, there is widespread interest in how the problem may be dealt with technically, politically, legally and financially. An increase in the demand for land, both for residential and commercial use, coupled with stricter planning controls on the development of green belt areas has led to an increase in the redevelopment of derelict land. Some of this land falls into the category of contaminated land. Alongside this generation of interest, several Government studies have been undertaken to examine the success of past policies and to explore some of the possible avenues for future policies. In 1985 the Royal Commission on Environmental Pollution (RCEP) recommended that the Government review the present system of registration of land charges and consider whether or not contaminated land should be entered as a charge. This marked the beginning of a series of related studies, culminating in the House of Commons Select Committee on the Environment publishing their report, "Contaminated Land", in January 1990. This report has had much influence on the changes which have occurred in contaminated land policy since then.

1.3 As guardian of the water environment the NRA has from the beginning been concerned about the pollution potential presented to both surface and groundwater by contaminated land. The NRA is equally aware of the importance of groundwater resources for human consumption, not least in the recent years of drought experienced in England and Wales, and has widely publicised its "Policy and Practice for the Protection of Groundwater" (NRA, 1992) document. It is therefore important to note that the NRA's interests lie in reducing the risk of contaminating groundwater, and surface water, and in prioritising the containment or remediation of such waters which are already contaminated. Such work has to be prioritised and this is best done in relation to risk, and its severity, relative to others within a particular catchment or overlying a particular aquifer. Prioritisation also has to be undertaken in relation to the NRA's duty to ensure that statutory water quality objectives are met. The NRA had therefore taken a keen interest in the developments surrounding the proposed registers of land subjected to a potentially contaminative use under Section 143 of the Environmental Protection Act 1990, and had submitted comments on the proposal. Now that the Government has withdrawn its proposals for establishing such a register, and is conducting a wide-ranging review of the subject in general, the NRA has drawn together a number of aspects relating to its own perspective of contaminated land as a contribution to this review. In doing so, however, the NRA has had to consider this as a long-term problem and thus of necessity has addressed possible solutions as being applicable to the proposed Environment Agency - as required in the Secretary of State's terms of reference for the review - rather than as being implementable by the NRA alone.

2 WHAT IS CONTAMINATED LAND?

2.1 Reflecting the wide range of interests in contaminated land, definitions vary considerably and it is difficult to pinpoint an exact definition suitable for universal application. Some land contains high natural levels of elements and compounds which may be regarded as contaminants in a general sense, but definitions usually relate to contamination as a result of human activity. The Department of the Environment (DoE) states that, at present, it is impossible to define contaminated land unambiguously and that contamination should be regarded as a general concept rather than something capable of exact definition or measurement. It does, however, define it loosely as:

"....land which represents an actual or potential hazard to health or the environment as a result of current or previous use" (DoE, 1989).

The DoE also cites a definition used by NATO's Committee on the Challenges of Modern Society, which defines contaminated land as:

"Land which contains substances which, when present in sufficient quantities or concentrations, are likely to cause harm, directly or indirectly, to man, the environment, or on occasion to other targets".

2.2 It is therefore useful to clarify the differences between contaminated and derelict land. Although the focus has been on contaminated land, there are links between it and derelict land, and the two often go hand in hand. Contaminated land is not exclusively derelict land however, because a site can be contaminated and yet still be fully operational. Equally, derelict land is not necessarily contaminated. The DoE defines derelict land as:

"....land so damaged by industrial or other development that it is incapable of beneficial use without treatment" (DoE, 1989).

2.3 The British Standards Institute (BSI) provides a different definition of contaminated land which clearly incorporates the concept that its use may be restricted by the presence of polluting materials:

"....land that contains any substance that when present in sufficient concentrations or amounts presents a hazard. The hazard may:

(a) be associated with the present status of the land;

(b) limit the future use of the land; and

(c) require the land to be specifically treated before particular use " (BSI, Draft for Development 175, 1989).

NRA Interest

2.4 The NRA has a specific interest in contaminated land because it may represent a source, or potential source, of water pollution. Contaminated land may cause contamination of both surface and groundwaters, but this in itself may not significantly affect water quality. It is only when contamination adversely affects water quality that a threshold is crossed; contamination then becomes pollution. In such a case, a controlled water will be polluted when either of the following factors are satisfied:

(i) that there is a breach of a water quality standard (WQS); or

(ii) that there is evidence of poor water quality or harm to the systems it supports.

It is also important at this stage to define the boundaries of NRA interest in contaminated land. The NRA is only concerned with a small proportion of the whole - specifically with that proportion of contaminated land which:

(a) is currently affecting surface water or groundwater quality in England and Wales or which is the sole cause of low river quality classification; and/or

(b) has the potential to pollute or downgrade controlled waters in the future, possibly as a result of its redevelopment.

2.5 There are thus two issues to be addressed. The first is to deal with controlled waters which are currently being polluted, and the second is to ensure adequate prevention of pollution of controlled waters in the future from contaminated land sources. This dual approach is reiterated throughout the report, and is seen as the logical approach to take. Specific NRA initiatives and proposals to manage this problem are dealt with in Chapter 8. However, prior to that, the foregoing chapters address apparent policy to date, relevant NRA legal powers, and the risks which are presented to the water environment from contaminated land sources. All of these factors influence the approach which may well have to be taken to manage the problem of contaminated land with respect to the water environment.

Contaminated Land - A brief history

2.6 Contaminated land arises largely as the result of past industrial processes which have left behind a legacy of many substances including oils and tars, "heavy" metals, organic compounds and soluble salts. Such land is largely situated in urban areas. The widespread mining for minerals and metals has, however, resulted in many rural and scenically beautiful areas of the country also being affected. But not all contaminated land is derelict, and thus although much of the contaminated land area is a result of the industrial revolution, many contaminative uses to which land may be put are very much industrial processes of today, as indicated in Table 1. Hence contaminated land has a history, but it is not a historical problem. This renders the management of contaminated land a twofold process: remedial works for land already contaminated, and the implementation of correct management and standards which will minimise future contamination.

TABLE 1 - USES OF LAND WHICH MAY GIVE RISE TO CONTAMINATION

(Source: Registers of Land which may be Contaminated: A Consultation Paper DoE, May 1991.)

Agriculture	Burial of diseased livestock.
Extractive Industry	Extracting, handling and storage of carbonaceous materials such as coal, lignite, petroleum, natural gas, or bituminous shale.
Energy Industry	Producing gas from coal, lignite, oil, or other carbonaceous material (other than from sewage or other waste) or from mixtures of those materials.
Production of Metals	Production, refining or recovery of metals by physical, chemical, thermal or electrolytic or other extraction processes.
Production of Non-Metals and their Products	Production or refining of non-metals by treatment of the ore.
Glass making and Ceramics	Manufacture of glass and products based on glass. Manufacture of ceramics and products based on ceramics, including glazes and vitreous enamel.
Production and Use of Chemicals	Production, refining, recovery or storage of petroleum or petrochemicals or their by-products, including tar and bitumen processes and manufacture of asphalt.
Engineering and Manufacturing Processes	Manufacture of metal goods, including mechanical engineering, industrial plant or steelwork, motor vehicles, ships, railway or tramway vehicles, aircraft, aerospace equipment or similar equipment. Storage, manufacture or testing of explosives, propellants, ordnance, small arms or ammunition. Manufacture and repair of electrical and electronic components and equipment.
Food Processing Industry	Manufacture of pet foods or animal feedstuffs. Processing of animal by-products (including rendering and maggot farming, but excluding slaughterhouses, butchering).
Paper, Pulp and Printing Industry	Making of paper pulp, paper or board, or paper or board products, including printing and de-inking.
Timber and Timber Products	Chemical treatment and coating of timber and timber products.
Textile Industry	Tanning, dressing, fellmongering or other processes for preparing, treating or working leather. Fulling, bleaching, dyeing or other textile floor coverings (including linoleum works).
Rubber Industry	Processing of natural or synthetic rubber (including tyre manufacture or retreading).
Infrastructure	Marshalling, dismantling, repairing or maintenance of railway rolling stock. Dismantling, repairing or maintenance of marine vessels, including hovercraft. Dismantling, repairing or maintenance of road transport or road haulage vehicles. Dismantling, repairing or maintenance of air or space transport systems.
Waste Disposal	Treating of sewage or other effluent. Storage, treatment or disposal of sludge including sludge from water treatment works. Treating, keeping, depositing or disposing of waste, including scrap (to include infilled canal basins, docks or rivercourses). Storage or disposal of radioactive material.
Miscellaneous	Premises housing dry cleaning operations. Laboratories for educational or research purposes. Demolition of buildings, plant or equipment used for any of the activities in the schedule.

3. CONTAMINATED LAND AS A PROBLEM - THE UK EXPERIENCE TO DATE

3.1 The UK has never actively sought to assess comprehensively the extent of contaminated land, although such programmes have been adopted in some other countries. In a House of Commons Select Committee report, published in 1990, specific reference was made to the reactive nature of national policy, in that it concerned itself almost exclusively with the end use of the land, and did not encompass the hazards to human health or the environment, whether potential or actual. There have been a number of isolated studies undertaken since the early 1970's, but most have attempted to assess specific categories of contaminated land, such as waste tips.

3.2 A test survey relating to potentially contaminated land was carried out in Wales in 1984, which was updated in 1988. The study was undertaken by Liverpool University and was jointly funded by the Welsh Office and the Welsh Development Agency. The aims of the study were to discover the scale of the problem in Wales and to classify sites according to:

(i) their potential hazard;

(ii) the need for remedial treatment; and

(iii) the factors which were likely to inhibit redevelopment.

3.3 In establishing the scope of the survey, two general exceptions were made: sites which were currently in use, and those sites of 0.5 hectare or less. In the case of potentially serious contamination, however, such as gasworks or tar lagoons, all sites were recorded because of their environmental significance. The results of the 1988 Welsh Office survey recorded 746 sites in total covering some 4,000 hectares (10,000 acres). Her Majesty's Inspectorate of Pollution (HMIP) also carried out a study in conjunction with Local Authorities, in a more limited way, in England and Wales to identify and locate waste disposal landfill sites generating gas; it was estimated that 1,400 such sites may have been emitting gas in quantities likely to present hazards.

3.4 The Department of the Environment (DoE), by extrapolating information from the most recent (1988) Derelict Land Survey, has estimated that a possible maximum of 27,000 hectares (67,000 acres) of derelict land could be classed as potentially contaminated in England (DoE, 1989). The figure amounts to 65% of the total derelict land, and 0.2% of the total land area. Although the DoE considered this estimate to be a maximum, it is noteworthy that it excluded land which was both in use *and* contaminated.

3.5 Other assessments have tried to include land which is currently in use. Environmental Resources Ltd (ERL) in 1987 estimated, from a pilot study carried out in Cheshire, that there were some 50,000 to 100,000 potentially contaminated sites in the UK. Other workers have similarly concluded that, based on European experience, there could be 50,000 suspect sites affecting as much as 50,000 hectares (120,000 acres) of land in the UK. However it was likely that only a small proportion would present an immediate threat to public health or the environment.

3.6 In attempting to estimate the amount of contaminated land, the problem of diffuse contamination must also not be excluded. Between 1979 and 1983 the Soil Survey and Land Research Centre undertook a National Soil Inventory of England and Wales, based on a 5 km survey grid, for some 13 substances. Taken as a whole, the soils were found not to be extensively contaminated with the elements investigated. Nonetheless, the study concluded that diffuse contamination was, if anything, more important than grossly contaminated industrial sites.

The British Geological Survey is currently undertaking for the Department of the Environment a reveiw of natural contamination in Great Britain and the results should be available late in 1994.

3.7 It might be thought that identifying the amount of contaminated, or potentially contaminated, land would assist in the development of positive and effective policies for dealing with it, but many European countries have already undertaken such studies and the daunting nature of the results may well explain why they have not yet taken this crucial first step. The Dutch inventory of contaminated land now stands at over 110,000 sites, Germany's at 100,000, and Denmark's and Finland's at 20,000 each. Britain's land use history would indicate that there are potentially 50,000 to 100,000 sites which could have been expected to be identified if national registers had been collated (ENDS Report 193, 1991).

3.8 In the UK plans to create, under the Environment Protection Act 1990 (EPA'90), a register of land which had been put to a contaminative use have now been withdrawn. Funds to deal with derelict land have however been available for more than 25 years as a result of the Local Government Act 1966, through a variety of legislation. These aspects are discussed in detail in later chapters of this report, but before doing so it is worth briefly reviewing some of the experiences and approaches in other countries.

Experience Abroad

3.9 The EC does not currently have a policy on the problem of contaminated land, but there are indications that steps are being taken in the direction of deriving such a policy. Such developments include the draft Directive on Landfill of Waste and the draft Directive on Civil Liability for Damage caused by Waste, and the Green Paper on Remedying Environmental Damage.

Contaminated Land in the Netherlands

3.10 The Netherlands has a high population density, high energy demands, and an intensive agricultural programme. Such factors make environmental pollution a risk which is mirrored in many developed countries, including the UK. However, the unique topography, hydrology and geology of the Netherlands result in a very high water table, which is constantly under pressure from high dependency on groundwater as a potable supply - in the region of 80% of national demands.

3.11 As with many environmental issues, the turning point in the Dutch approach to contaminated land came only after disaster had struck. In 1980, 1,600 drums of illegally dumped toxic waste were discovered in a small town called Lekkerkerk. Several hundred houses had been built on the site, which was a reclaimed waste tip. Widespread illness affected local residents and on inspection it was discovered that both the groundwater and underfloor voids had been contaminated with toluene, and a variety of other organic chemicals. The Dutch Government funded the clean-up operation which, in 1981, cost £156 million. The site was successfully decontaminated, but almost twelve years later the transport company responsible is still being pursued through the courts.

3.12 This event led to a widespread and intensive approach to cleaning up contaminated land. The Dutch policy has been based on three elements:

(i) an integrated environmental approach;

(ii) *in situ* treatment wherever possible; and

(iii) the concept of multifunctionality - essentially a use-related practice so that, when remedial works are complete, any use may be made of the land.

The concept of multifunctionality is not without its critics and the cost effectiveness of cleaning up each site to such quality must be very seriously questioned.

The USA and Superfund

3.13 As with the Netherlands, it was an accident that prompted action on contaminated land in the USA. Again the cause was a waste disposal site which had been capped, and despite advice not to redevelop it, a school and residential development were built in the late 1950's. By 1976, unexplained illnesses were reported by residents and an investigation began. The site, known as Love Canal, was declared a national emergency following initial investigations, and a State sponsored remedial programme was launched immediately. The incident resulted in Superfund. Under the Comprehensive Environmental Response, Compensation and Liability Act 1980 (CERCLA), the US Congress established Superfund at $1.6 billion, increasing to $8 billion in 1986. The majority of the funding was obtained from a levy put on the manufacture and import of certain chemicals, the remainder came from general taxation.

3.14 The US Environmental Protection Agency's (US EPA's) Fiscal Superfund report 1991 stated that all surface contamination had been removed from 196 of the 600 priority sites, and final clean-up was achieved at 63 sites. The programme is to have around 200 sites ameliorated by early 1994, and 650 by the year 2000.

3.15 It is clear that Superfund, as a concept, is an interesting approach to contaminated land because it attempts to follow the "Polluter Pays Principle" and appears to be having an impact on industry, in that it is providing incentives. It would appear that the principal problems lie in the administration of the programme, and the standards set, and it has been extensively criticised for the excessive bureaucratic delays which exist, which are claimed to result in unnecessary wastage of funds. The net result is that less sites are now being classified and treated. There is also a plethora of litigation to determine the party or parties responsible for the contamination of each site (Kreager D, 1991). The USA, Dutch and some other policies which have been adopted are indicated in Table 2.

TABLE 2: EXAMPLES OF CLEAN-UP POLICY IN FOUR COUNTRIES

Source: (Contaminated Land: Market and Technology Issues, CEST 1992.)

COUNTRY	LEGISLATION	GOALS	FINANCES	METHODS
USA	CERCLA 1980 (amended by SARA 1986).	Clean-up of abandoned and uncontrolled hazardous waste sites.	Enforcement of responsible parties or central funding.	Hazard ranking system; National Contingency Plan.
Netherlands	Interim Soil Clean Up Act 1981 (Soil Protection Act).	Restoration of soil to condition suitable for any given use.	Mainly Government, but also responsible parties.	Guideline for Soil Protection and Regulation on Building Materials.
Germany	Expected in new Soil Protection Act.	Restoration of soil with respect to the proposed use of land.	Responsible parties and Government.	Use of newly developed guidelines being prepared by Ministry of Environment.
Denmark	Waste Disposal Sites Act (revised 1990)	Groundwater protection as a high priority.	Mainly Government, but also responsible parties.	Guidelines on setting of priorities on soil quality criteria, and the reuse of soil, are in use or discussion.

4. WATER POLLUTION - ACTUAL AND POTENTIAL

4.1 The NRA is committed to the protection of both surface and groundwater resources from contaminated land sites. Not only is there a statutory duty to do so under national and international legislation, there is also a requirement to maintain and where necessary improve the quality of water with respect to its sustainable use. When approaching the problems associated with contaminated land, the question which the NRA therefore needs to ask is - "What are the risks to the water environment?"

4.2 This is a difficult question to answer accurately for a number of reasons. Firstly, two of the most important indicators - the source and the receiving water - are often hidden from view and therefore quantifying or carrying out a risk assessment is a major operational undertaking. Similarly, because of other pollution sources - such as consented discharges, naturally-occurring sources of metals, or pollution incidents - attempts to quantify the percentage of pollution in a watercourse attributable to a contaminated land source is equally difficult. Seasonal and climatic fluctuations in the water cycle further complicate the situation. And although the industrial practices of 50 or more years ago may be known, the exact amounts of contaminants which may have been released into the environment are generally unknown. Similarly, because much contamination will take place over a long period of time, environmental monitoring will often only begin to show adverse patterns once pollution has already occurred. Thus the majority of information on contaminated land and its impact on the water environment tends to be site specific and only a few general rules may be assumed. In addition, because policy has been focused mostly on redevelopment and end use, this is the aspect upon which most knowledge has been accumulated. What can be ascertained without actually sampling or monitoring can only be a general risk assessment based on local environmental factors. Some more specific risks are cited later in this chapter, where pollution has actually occurred and monitoring networks have been established; however, to begin with, some general risks to the water environment need to be considered.

General risks

4.3 Contaminated sites such as old gas works, landfill sites, chemical works and heavy engineering sites can present a risk to either groundwater and/or surface water depending on the local geology and hydrogeology. This applies to sites in an active, dormant, or derelict state and more especially as a result of disturbance during redevelopment. Natural or artificial seals, such as old concrete foundations and basement floors, clay layers, or other natural foundations which have become clogged by seepage of contaminants, may exist beneath the site. Based on the geology of the area an assessment of the likelihood of leaching can often be made. Permeable rocks will obviously be more vulnerable to pollution than rocks such as clays and shales, which will inhibit percolation downwards to groundwater, but promote run-off to surface water. It is therefore necessary for each site to have its own assessment in order to establish a comprehensive picture of the risks involved: factors such as site history, local climatic conditions, any proposals to disturb or remove the soil, local water courses and their uses - all need to be included to achieve an overall measure of the risk.

General Risks - Redevelopment of Contaminated Land Sites

4.4 The redevelopment of contaminated or derelict land sites has provided most information about the risks to the water environment. In many cases this has been the first opportunity for the NRA to become aware of, and involved in, investigatory, remedial or preventive work. There are a few rules of thumb which can be followed, but again an assessment of each site is the most accurate way to assess risk and each site will vary according to local conditions.

4.5 Disturbance caused by site investigation boreholes, site demolition or stripping, and construction of foundations, can all cause the release of contaminants into an underlying aquifer. The removal of surface cover (roofs, floors) opens the site to rainfall which can cause pollutants to leach to either ground or surface waters.

4.6 Piling in contaminated ground can enhance the downward migration of contaminants both during and after construction. Vibro-replacement piles, where numerous reinforcing columns of compacted granular material are introduced into the ground, can be hazardous to the underlying groundwater, by allowing a direct flow down the column into the underlying aquifer.

4.7 New technology for the treatment of contaminated ground, such as the use of microbiological degradation, may also present a short term risk to groundwater quality if adequate precautions are not taken. The precaution of lining a site would minimise this risk. Other novel forms of chemical, physical and thermal treatment are still largely unproven.

4.8 Where removal of contaminated material takes place, this can expose relatively fresh pockets of contamination which may be susceptible to oxidation and leaching. Heavy rainfall during such operations can result in pollutants entering either surface water systems or underlying aquifers. Where development plans include the creation of landscaped areas, contaminants present in the ground may enter the groundwater or surface water systems unless steps are taken to minimise infiltration, such as installing a clay seal.

4.9 The provision of groundwater quality monitoring boreholes around the edge of a site is usually requested as a condition of planning permission being granted, and is often provided voluntarily by the developer. However, the groundwater is rarely monitored long enough to allow the effects of the redevelopment to be adequately assessed by the NRA.

Examples of Actual pollution

4.10 To enable the NRA to make a comprehensive statement on the threat which contaminated land sources present to the water environment, the NRA has undertaken a preliminary study of such sites in England and Wales. This revealed far more than mere numbers: it also served to illustrate the variety of sites which are actually causing, or have the potential to cause, pollution. No sampling was done specifically for the study. It was based on the knowledge and involvement of experienced pollution control officers to identify areas of contaminated land which were adversely affecting the water environment

4.11 In order to illustrate the types of pollutants that can arise in both surface and groundwaters, from contaminated land sources, a few examples are given below. All have been investigated by the NRA and are sites which have crossed the threshold from contamination to pollution: all have either breached standards or represent proof of actual deterioration in water quality. Some are derelict, some are operational. It is important to note that in some cases remedial action is taking place, or is planned. The NRA has been actively involved in the negotiations to undertake such works.

Landfill sites

4.12 Completed landfill sites are one of the more obvious and frequently cited forms of contaminated land. Landfill practices in the past have led to many instances where leachate has contaminated soil, ground and surface waters. Again, problems experienced tend to be site specific, but the typical array of contaminants that can be, and have been, found are summarised below.

4.13 A completed landfill site in Northumbria/Yorkshire Region had been used, a number of years ago, for the disposal of cyanide containing substances (Figure 1). The resultant leachate caused contamination of the groundwater to levels above those specified in the EC Drinking Water Directive. One observation borehole recorded levels three times the specified limit. The owner agreed to remove the material, but concerns over health and safety during the removal operation initially delayed the start of the remedial work. The removal work is now well under way and is being carried out by the County Council, but is proving expensive and other methods of reducing leachate flow are being considered (Figure 2). As can be seen, above ground appearances can be deceptive.

4.14 In another example, this time in the North West Region, significant levels of Polychlorinated Biphenyls (PCBs) have been found in an old waste tip which, licensed since 1970, has been used for tipping industrial wastes, petrochemical sludges, redundant transformers and oil (Figure 3). The site is currently derelict. It is underlain by glacial boulder clay of up to 40 m in depth, which in turn lies over a sandstone aquifer. Significant levels of PCBs have been found to be present below the point of discharge in the local watercourse (Figure 4). The brook is 0.5 km away from the site and measures 4 km in length. It is polluted with PCBs and is unclassified. The NRA has been involved with this site for some time and currently undertakes bimonthly sampling of the watercourse. It is also carrying out an ongoing study of the biological conditions in the watercourse. A geological survey has been carried out. The NRA has been in discussion with the owner of the site for some time and has given its support to his efforts to obtain financial assistance to clean-up the site both from the companies responsible for tipping the wastes and via the Derelict Land Grant (DLG). The estimated cost of clean-up is £2 million. The application for DLG was turned down apparently on the grounds that the tip itself is not derelict and other sites had priority. Other sources of funding are now being pursued and legal action has been initiated by the NRA.

4.15 In the NRA's Southern Region an old sand quarry which has recently been landfilled is causing water quality problems. The site has now been capped and restored to agricultural use. A liner was installed, but has subsequently been found to be ineffective, with some perimeter monitoring boreholes showing elevated concentrations of ammonia (20 mg/l) and chlorides (500 mg/l). This is causing concern because of the need to maintain water quality in a public water supply borehole 0.5 km away. There is also concern over the water quality in a nearby watercourse, which is only 100 m from the site. The site operator has agreed to remove increased quantities of leachate from the site in an effort to improve the quality of the surrounding aquatic environment. The NRA is closely monitoring the current situation to ascertain the extent of any environmental improvement.

4.16 The historic landfilling of waste is also causing water quality problems in the South Western Region. A site used to dump greasecake from the wool scouring industry has since been redeveloped as an industrial estate (Figure 5). Prior to development of the site remedial measures were undertaken, with all heavily contaminated areas being removed and disposed of to a suitably licenced landfill site. Less contaminated areas were capped with clay, or paved, and the surface water drainage from impermeable areas piped to stream rather than to soakaway. The River Tone flows adjacent to the site and analyses have shown elevated levels of permethrin; there are also detectable levels of dieldrin and lindane, but these are within environmental quality standards. Biological surveys have been carried out and indicate that the macro-invertebrate community of the stream downstream of the site is impoverished and less diverse than would be expected; however, contributions from other sources cannot be ruled out.

General Industry

4.17 A range of heavy industrial activities has presented the NRA and its predecessor bodies with a variety of water quality problems. A selection is summarised below, which includes historic sites which have long since discontinued activity as well as a number of fully operational ones.

4.18 An industrial estate in the Thames Region has a partly developed complex which includes the storage of trichloroethylene. The storage facility originally used open tanks, and as a result of a spillage from the tanks the local groundwater became heavily contaminated. Trichloroethylene is now stored in closed storage structures with appropriate bunds. The site is within 1 km of the River Thames and the nearest public supply borehole is only 1.5 km away and remains - to date - unaffected. However, the potential for contamination remains; the site is underlain by a shallow gravel aquifer in which trichloroethylene has been found in concentrations of the order of 200 mg/l. The owners of the site have agreed to implement a detailed programme but as yet work has not started. The NRA is still pursuing the recovery of the expenditure it made during its initial investigations.

4.19 Also in the Thames Region is a brick manufacturing industry which now occupies a site which was previously used for chemical manufacturing during the Second World War. The industrial processes included the production of sulphur, for vulcanised rubber, and the manufacture of zinc and bromate salts. As a result of these activities the area around the plant has become contaminated with zinc, ammonia, bromide and bromate. There are localised areas of very low and very high pH. Part of the site was also used for waste disposal, although there are no records of what was tipped or the extent of the tip area. The main impact of this site, however, is on the surface waters: one unclassified watercourse and one National Water Council (NWC) Class 2 river. There is evidence of some contamination of the surface water resulting in a failure of the historic RQOs (non-statutory River Quality Objectives) and there is also an impact upon a standby water abstraction point. The site does not cause serious pollution because the effluent is collected in a series of lagoons and treated prior to discharge.

4.20 Elsewhere in the Thames Region a former British Rail Engineering works near Swindon is creating a number of problems to local water quality. The site incorporated extensive heavy engineering involving the manufacture and repair of locomotives; it also had its own gas works. It has been partially redeveloped, with further phases ongoing. The River Ray, a NWC Class 2 watercourse, flows through the site. Evidence of pollution has been found in the underlying aquifer; there is also contaminated water flowing across parts of the site. There is potential for contaminated water to enter the watercourse. As a result of the extensive size of the site and its piecemeal development, site investigations have been sparse. A number of hot spots of contamination have been identified and proposals put forward to remove them. Relationships with the developer have improved significantly of late and as a consequence those parts of the site which have recently been redeveloped have been done so in such a way as to minimise the impact on the water environment. However, there are areas which will remain in their current state and others which were redeveloped prior to NRA involvement where risks from contamination remain.

4.21 In the South Western Region a derelict site, long used for the extraction of heavy metal ores including tin and copper, and for the refining of arsenic has the potential to cause serious water pollution to the local watercourse. The site contains many unvegetated spoil tips, (Figure 6) some of which are actively eroding or have a high risk of collapse. The arsenic content of the material on the site can exceed 5% and copper can exceed 1%. Streams draining the site (Figure 7) frequently have concentrations exceeding 35µg/l of arsenic and 400µg/l of copper. There is a commercial forest adjacent to the site (Figure 8) where felling in future years is likely to increase the potential for erosion. There is also a potential for gross pollution of the River Tamar in the event of underground or surface collapse. If this should happen, it could have serious implications for the public potable supply abstraction situated 2 km downstream of the site.

4.22 A small laundry with a dry cleaning operation has been one of several sites causing pollution in the Anglian Region. The laundry has been in existence for many years, during which time inadequate disposal practices have resulted in groundwater contamination with chlorinated solvents. The site is located on gravels which overlie the unconfined Middle Chalk and is close to the Rivers Linnet and Lark. There is a very large potable source abstraction nearby. There has been extensive analysis of boreholes in the area and a number of supplies are known to be affected. In one borehole 1.5 km away, chlorinated solvents have been found at 0.5µg/l, whilst at a local brewery, 0.75 km away, the concentration is 40µg/l; perchlorethylene concentrations of up to 1500µg/l have been found in boreholes in the area. The site owners, after consultation with the NRA and other specialists agreed to implement a remedial programme of excavating and removing the most heavily contaminated ground. This is now complete and initial indications are favourable, although the NRA is monitoring the position closely.

4.23 An area of considerable concern is that of a ship breaking yard in Northumbria/Yorkshire Region which has caused extensive PCB pollution at Battleship Wharfe, Blyth Harbour (Figures 9 & 10). The NRA has been involved with the site for some time. The wharf operator was breaking up two Russian vessels when an oil spillage occurred. The NRA was notified and the spillage contained. Following this, the operator went bankrupt and another operator was brought in to finish scrapping the two vessels. The wharf is now empty. Investigations found PCB contaminated sediments on the bed of the wharf and in the main river with a maximum concentration of 35.3 mg/kg (wet weight). Since this discovery, the NRA and MAFF have been involved in a number of sampling and monitoring exercises. Remediation costs are estimated to be in the region of £500,000. Discussions are presently continuing over the viability and costs associated with the cleaning up of the contaminated sediments within the harbour and the estuary. Also, the clean up of an area of contaminated land adjacent to the harbour is being considered. While the way forward is, at present, unclear there is no doubt that both the Harbour Authority and NRA desire environmental improvements within the area.

4.24 Some sites present a mixture of problems. For example, a smelter site in the South Western Region covering some 400 hectares is known to have been causing widespread contamination to local surface and groundwaters. The site is partially derelict, partially developed, with plans for further development. Some disposal of slag from the smelter was carried out within the site; some slag is stockpiled. Parts of the site have been raised and slag waste in the form of smelting briquettes is distributed across the area (Figure 11). The site has also been used for the disposal of other chemical manufacturing industrial residues. The main contaminants are lead, zinc, cadmium and other metals associated with the smelting processes. The site is adjacent to the Severn Estuary into which surface waters discharge (Figure 12). The NRA is currently investigating discharges from the site and in particular the contributions from point sources from the contaminated land. No proposals from the site owners have as yet been put forward for remediation work; these will be fully explored by the NRA once the full extent of the problem is known.

4.25 In the Severn Trent Region a tipping site close to a copper refinery has given rise to a number of problems. The site itself is located over old mine workings and is currently derelict. It is underlain by shales, mudstones, coal measures and fissured sandstones. The underlying coal workings appear to be saturated to the surface and the water table is at ground level in places. The River Tame passes adjacent to the site. There is evidence of pollution of the river from contaminated groundwater seepages and run off. The NRA has carried out extensive monitoring of the river and a site investigation which was jointly funded between the NRA and the site owner. Elevated levels of copper, nickel, zinc and iron have been found. It is considered that contamination in the vicinity of the refinery (operating since 1917) has polluted groundwater and the contaminants have subsequently migrated so that, today, the quality of the River Tame is being adversely affected (Figures 13 and 14). Investigations are ongoing to quantify the extent of this complex problem with a view to producing a strategy for the implementation of remedial measures.

4.26 Severe contamination of chalk groundwater has been discovered in Southern region beneath a site previously used for the manufacture of tar based insecticides. Extremely high concentrations of phenols (up to 28 mg/l) and polycyclic aromatic hydrocarbons (up to 1.7 mg/l) have been observed in groundwater samples obtained from investigatory boreholes, and a floating layer of "creosote" is present on the water table. A clean-up strategy has now been agreed with the current owner and Local Authority, as a condition of the planning consent for redevelopment.

Agriculture and Horticulture

4.27 Contaminated land is generally associated with heavy industrial activity; however, the NRA has been involved with sites where contamination originates from the agricultural and horticultural industries. The first example illustrates that, in some cases, metal contamination can occur from natural sources. Drainage of peat soils overlying metal rich sub-soil in the Brightley Stream catchment in the South Western Region has produced significant water quality problems. Exceptionally high levels of metals are mobilised from various mineral sulphides in the soil which gain easy access to the stream via land drains as shown in Table 3. The problem is particularly severe after long periods of dry weather. Acidophilic bacteria cause a build up of both high metal concentrations and very acid conditions during dry weather. During the first flush of heavy rain, poor quality water is leached from the soil and this has contributed to fish kills in 1976, 1978, 1984 and 1989.

TABLE 3 WATER QUALITY OF BRIGHTLEY STREAM DURING HEAVY RAINFALL, OCTOBER 1989, (FROM NRA WATER QUALITY SERIES NO 6 1992)

Concentrations expressed as milligrams per litre (except pH).

Determinand *	Land Drain	Brightley Stream downstream of land drain
pH	2.7	3.1
Dissolved sulphate	942	250
Total aluminium	267	6.51
Total nickel	2.36	0.34
Total copper	0.68	0.13
Total cadmium	0.11	0.02
Total zinc	33	4.97
Total iron	130	29

4.28 The NRA South Western Region has also had serious pollution problems from another unconventional source of contaminated land - daffodil farms. Historically, much of West Cornwall has been used for daffodil bulb and flower production (Figure 15) as well as for potatoes and brassicas. Aldrin was applied to daffodil bulbs to control the large narcissus fly and, on land converted from old grass to potatoes, to control wireworm. Both aldrin and dieldrin were used on the brassicas to control root fly. Both compounds adhere to organic soil particles which are flushed into the rivers during rainfall. Cultivation to a fine tilth, weed control, and a high level of both wheeled and foot traffic in daffodil fields all encouraged the formation of colluvium during rainfall, and the colluvium contained high levels of pesticide residues. A comprehensive investigation was undertaken by the NRA South West and a report submitted to the Advisory Committee of Pesticides. This resulted in the withdrawal of the approved use of aldrin on daffodils throughout the UK.

Mine Spoil Tips

4.29 The problem of abandoned mines warrants a separate study; however, associated spoil tips can be categorised as a contaminative land issue. Although many examples may be cited, only one will suffice as being representative of the nature of the problems which such tips present. In Severn Trent Region the spoil tip of a lead mine which operated between 1866 and 1921 remains derelict and unvegetated. The site has also been used to dispose of old gas works' waste in small isolated pockets. The Afon Cerist flows through the site and a brick lined drainage culvert receives run-off from the tip which flows into the river and subsequently changes it from NWC Class 1A to Class 4.

The river remains contaminated with lead and zinc for ten kilometres downstream. Elevated amounts of other trace metals are also present in the river. There have been several attempts by developers to reclaim the site and a scheme is currently underway. It is being undertaken by Powys County Council with 100% funding from the Welsh Development Agency and is due for completion in the spring of 1994. The NRA is monitoring the metal levels in the river to establish whether the scheme results in significant environmental benefits.

A recently completed study for the Department of the Environment has reviewed existing treatment methods for metalliferous mining sites, including taking account of the water environment. It sets out recommendations for site assessment, reclamation and management. It provides a guidance framework on the best practicable options for dealing with sites contaminated as a result of previous mineral workings, and also on the prevention of contamination at new workings. (University of Sheffield, 1994)

5. THE LAW

Statutory Duties and Powers of the NRA

5.1 The NRA has a number of legal powers to help it prevent, manage, and ameliorate a variety of water quality problems, including those associated with contaminated land. They arise from Water Resources Act 1991 (WRA'91), from the Environmental Protection Act 1990 (EPA'90) and the Town and Country Planning Act 1990 (T&CPA'90)(as amended by the Planning and Compensation Act 1991) (P&CA'91). Local Planning Authorities are required to consult the NRA when preparing the proposals for a statutory plan or for the alteration or replacement of such a plan and before finally determining the contents of the proposals. Statutory plans in this regard means unitary development plans, structure plans, local plans, minerals local plans, waste local plans.

Local Planning Authorities are also required to consult the NRA before granting permission for development which falls within categories set out in the General Development Order 1988 (GDO). Operations where an LPA must consult with the NRA include development involving mining operations, the use of land for the deposit of waste, for the purpose of refining or storing mineral oils, or for carrying out works or operations in the bed of or on the banks of a river or stream.

In addition to these requirements the Secretary of State has requested Local Planning Authorities to consult with the NRA on certain other development proposals (which could lead to increased industrial discharge to a river or estuary or which will lead to increased drainage problems in areas notified as having high water tables). There is however no obligation on an LPA to consult in these circumstances. Together these powers should provide access to the three main avenues of managing water quality problems resulting from contaminated land sites:

i) stopping current adverse practices;

ii) cleaning-up problem sites where possible; and

iii) preventing future problems by ensuring better management practices.

They may be seen as being capable of use independently of each other, or in combinations. These avenues and their related legal powers are summarised below.

TABLE 4 - MANAGEMENT STEPS AND ASSOCIATED LEGAL POWERS

STOPPING CURRENT POLLUTION SOURCES	CLEANING-UP PROBLEM SITES	PREVENTING POLLUTION IN THE FUTURE
OPERATIONAL SITES	OPERATIONAL AND/OR DERELICT SITES	REDEVELOPMENT OF DERELICT SITE AND PROPOSED NEW INDUSTRIAL SITES
WRA'91 - Section 85 - Section 86	WRA'91 - Section 161	WRA'91 - Section 92 (feasibility to be considered) - EPA'90: - PLANNING ACTS:
DIRECT NRA POWERS/ACTION	DIRECT NRA POWERS/ACTION	CONSULTEE STATUS

5.2 It is worth recording, at this point, the NRA's relevant duties with respect to water quality. A basic function of the NRA, as set out in Section 3(2) of the WRA'91, is to protect against pollution any water which is, or is likely to be, abstracted; and Section 19 places a duty on the NRA to take all such actions, as it may from time to time consider necessary or expedient, for the purpose of conserving and securing the proper use of water resources. More specifically, in relation to water quality, the NRA is under a duty to achieve - with the provisions of the pollution control chapters of the Act - any Water Quality Objective (as determined by reference to relevant standards) set under Section 83 of the WRA'91. In order to achieve such objectives, the NRA has powers to control the direct discharge of effluents into water. It also has powers to prevent and ameliorate pollution.

5.3 Chapter II, Part III of the WRA'91 contains the principal water pollution offences. Under Section 85(1) of the Act, it is an offence if a person:

"…. causes or knowingly permits any poisonous, noxious or polluting matter or any solid waste matter to enter into controlled waters …."

unless that entry is covered by a consent to discharge issued under Section 88 of the Act or has a defence under other provisions in that Section and in Section 89. A person who contravenes this section of the Act is liable to be prosecuted, which could result in a fine, imprisonment, or both.

5.4 Section 86 of the WRA'91 provides the NRA with a legal power to prevent certain discharges from taking place, in the interests of protecting the water environment, by serving a notice, known as a "prohibition notice", on a person or persons in conjunction with the contravention of Section 85(4) (see Appendix 1). This preventative power is limited to that over the discharge of trade or sewage effluent to land (ie soakaways) or of discharges, other than trade or sewage effluent, from drains or sewers. A prohibition notice may require the discharge to cease or may allow for the discharge to be continued in accordance with a number of conditions set out in the notice. Failure to comply with a prohibition notice can result in prosecution under Section 85 of the WRA'91, as a principal pollution offence. However, a prosecution may only be taken while the notice is operational - which is at least three months from the date the notice was served. As has been stated by the NRA in its Groundwater Protection Policy (NRA, 1992), it will continue to use these powers, not only in respect to groundwater resources, but where surface water pollution is occurring.

Direct Powers - (i) Ameliorating Water Pollution

5.5 Not all cases of water pollution, such as many of those resulting from a contaminated land site, can simply be stopped at source. Many problems are historic in nature and pollutants may well have been released over long periods of time; this may render identification of both the pollution and its source a lengthy process. Often in such cases a clean-up operation will be required and the NRA has direct powers under Section 161 of the WRA'91 to undertake clean-up projects, and to recover the costs. The section covers anti-pollution works and operations (see Appendix 1) and enables the NRA to take steps to remedy any pollution affecting controlled waters. The section also entitles the NRA to undertake works or operations in order to prevent such pollution. Once either of these actions has been taken by the NRA, it may then seek to recover its costs from the person(s) who caused or knowingly permitted the pollution to take place, with the exception of pollution resulting from waters permitted to flow from abandoned mines. Even where a company has disposed of the land to another party, it may still be called upon to contribute towards the cost of the clean-up operation. The definition of a "person responsible" may also go much further than simply the owner of the land; anybody having an active role in managing the land could well be implicated, such as mortgagees, landlords and insolvency practitioners (Bryce A, 1992), but this would depend on the interpretation by the courts as to who *caused* or *knowingly permitted* the material to enter the controlled waters. Under Section 162 of the WRA'91, the NRA may also carry out works on land to ensure that water "…… in any relevant waterworks …." is not polluted or otherwise contaminated. Such relevant waterworks may include boreholes, springs, wells, adits or even reservoirs. The cost of carrying out such works, however, cannot be recovered under this section of the Act.

Direct Powers - (ii) Prevention of Future Problems

5.6 Section 92 of the WRA'91 (formerly section 110 of the WA'89) states that the Secretary of State may, by means of regulations, make provisions requiring a person who has custody or control of poisonous, noxious or polluting matter to carry out prescribed works and precautions to prevent or control the entry of matter into any controlled waters (see Appendix 1). This Section of the Act has so far only been used to introduce The Control of Pollution (Silage, Slurry and Agricultural Fuel Oil) Regulations 1991, which came into force on 1 September 1991. In addition, Section 92 allows for the regulations to confer power on the NRA to enable it to determine and specify what needs to be done (according to a particular situation relating to those Regulations), and if such specifications are not undertaken, to take legal action. Essentially such powers - none of which have been granted - could amount to what might be termed improvement notices, which would enable the NRA to require specific action to be taken, by a certain date.

Discussion

5.7 The NRA has considerable powers to prosecute when pollution is caused by whatever means, including that caused by the redevelopment of contaminated land, and of ameliorating pollution. Considerable as the direct powers available to the NRA are, however, there are some practical problems associated with them. In the case of Section 161, it is improbable that this section could be easily used for large areas of contaminated land, given the expenditure associated with clean-up programmes. It is, in any case, unlikely that the Section was intended to provide more than a 'fire fighting' power, the NRA being primarily a regulatory body with respect to pollution control. And one of the main difficulties in relation to pursuing and securing a prosecution under Section 85, or of recovering its costs - with or without legal action - under Section 161, is that of identification of the polluter, the owner, or person otherwise responsible. Contamination may have been occurring slowly and unobtrusively for many years, in some cases for decades, before a deterioration in water quality became apparent. In such cases it is unlikely, even in the event of identifying the original owner or party responsible, that they would be able to contribute to all or some of the costs of the clean-up operation; these are liable to be very expensive. It is also the case that, even if cost recovery was guaranteed, the NRA would be unlikely to have sufficient capital to begin remedial projects of a large scale. For example, if only 10% of the known contaminated land sites affecting water quality in the NRA's Severn Trent Region were to be cleaned up under Section 161, assuming a modest cost estimate of £0.5 million per site, the total is of the order of £10 million. It is therefore highly unlikely that even the principal clean-up operations concerning the NRA's statutory duties to protect the water environment could be achieved through Section 85 or Section 161 of the WRA'91.

Of greater relevance, however, is the fact that schemes required are not usually those which can be solved simply by the injection of capital money. The majority of solutions require systems which then need to be run; this would require the NRA to be a long-term operator of treatment processes rather than a regulator of the discharges of others. In essence, therefore, the issues are as follows:

- it is difficult to identify effectively the persons directly responsible;

- the costs of clean-up are large (and hence it is difficult to find landowners readily able to pay); and

- the NRA itself simply does not have, nor should it necessarily be given, the resources even to begin to tackle significant clean-up projects (even if it could recover the costs later); because

- it would be inappropriate for the NRA to become an operator rather than a regulator with respect to this cause of water pollution.

Indirect role of the NRA

5.8 In addition to its own powers, the NRA has an indirect influence on the management of contaminated land sites via the EPA'90 and the planning process. The planning process in the UK has until recently been the principal means for initiating remedial action where necessary. Essentially this has resulted in the issue of remedial work being addressed by means of planning conditions attached to the granting of planning permission. It is fair to say that, up until now, the whole management of contaminated land has been almost exclusively concerned with the end use of land and has hence only arisen in the event of an application for the redevelopment of a site being submitted to a local planning authority.

The Environmental Protection Act 1990 (EPA'90)

5.9 The introduction of the Environmental Protection Act 1990 now provides additional scope for the NRA's knowledge of the water environment to be called upon for a number of environmental management issues, including contaminated land.

5.10 The EPA'90 aims to manage, improve, and protect the natural environment in the UK. The Act has a number of sections which are specifically related to both the clean-up of land currently contaminated and the prevention of future contamination through better management practices. The most relevant chapters of the Act are Parts II (Waste on Land) and VIII (Miscellaneous). As has been identified in a pilot study undertaken by the NRA's Severn Trent Region, landfill sites represent a large proportion of contaminated land categories which are having adverse impacts on the water environment in some areas. The provisions of the EPA Parts II and VIII are, therefore, of vital importance to the NRA because proper enforcement in the future will have a large effect on this particular category of land.

Section 33 and 36: Treatment or Disposal of Waste, and Licensing

5.11 Section 33 of the EPA'90 sets down the foundations of the waste management licensing system by prohibiting the deposit, treatment, keeping or disposal of waste in or on land, or by means of a mobile plant, except under and in accordance with a waste management licence. Section 33(1)(c) creates one of the main disposal offences under Part II, namely not to "treat, keep or dispose of controlled waste in a manner likely to cause pollution of the environment or harm to human health". Controlled waste is defined in Section 75 of the Act while Section 36 deals with the granting of waste management licences. Section 36(4) imposes a duty on the Waste Regulation Authority (WRA) to consult the NRA before it issues any licence. If the NRA requests that the licence is not issued or disagrees with the conditions of the licence, and the disagreement cannot be resolved, the licence may not be issued except in accordance with a decision made by the Secretary of State under Section 36(5) of the EPA'90. Essentially, any decision taken on a licence, including its surrender, must first be approved by the NRA. Being involved at the licensing stage of waste management will help the NRA to ensure that precautionary and preventative measures are used where potential for water pollution may arise. The Part II licensing provisions are expected to be implemented during 1994.

Section 34: The Duty of Care

5.12 The EPA'90 also imposes a *duty of care* on any person who "....imports, produces, carries, keeps, treats or disposes of controlled waste, or as a broker, has control of such waste..." The duty extends to preventing illegal deposits of waste and the escape of waste from a person's control and to securing that the waste is transferred to an authorised person. It also requires a written description of the waste and other details in the form of a transfer note which must be passed on to the person who receives waste. The duty applies to both a person receiving waste and a person passing on waste to another party. The implications of this section are that companies will have to develop new contractual procedures to ensure that there is no breach of this duty. A breach of the criminal law may also give rise to civil proceedings if significant losses are incurred by a third party. A wide range of people will be regarded as "keepers" of waste and it remains to be seen if this expression will be extended to persons in control of contaminated land who do not take steps to control it.

Section 61: Clean-up Liability

5.13 Section 61 of the EPA'90 places a specific duty upon each WRA to inspect its area to detect whether any land is in such a condition by "relevant matters" affecting the land, that it might cause pollution of the environment or harm to human health. The "relevant matters" to be addressed are "... the concentration or accumulation in and emission or discharge from the land of noxious gases or noxious *liquids* caused by the deposit of controlled waste in the land...."

5.14 WRAs under the EPA'90 may enter land at any time where they have reason to believe that there may be concentrations of liquids or gases which might cause pollution. In the event of this being confirmed, the WRA then has a duty to take steps as appear to it reasonable to avoid pollution. The costs of such works may then be recovered from the "....person who is for the time being the *owner of* the land, except such of the cost as that person shows was incurred unreasonably....". "Owner" is not defined by the Act; however, under the Planning Acts and other legislation, the owner is defined as the person who whether in his own right or as a trustee for any other person, is entitled to receive rent from the land, or would be if the land was let. It should be noted that unlike Section 161 of the WRA'91, the original disposer has *no* liability under Section 61 of the EPA'90 unless he is still the owner. Again, as with Section 161 of the WRA'91, the question of the financial ability of the relevant Authority to undertake remedial work will be an issue, and the familiar scenario of "no guarantee" of cost recovery could well have a negative impact on the use of these powers by WRAs.

5.15 Once Section 61 has been implemented, the following will apply. Before a waste management licence can be surrendered, the WRA must be satisfied that no future pollution of the environment or harm to human health is likely to result. In considering whether to accept the surrender of the licence, the WRA has to consult the NRA to ensure that they are content; if there is disagreement, the matter must be referred to the Secretary of State. An additional duty is placed on each WRA, under Section 61(5) of the EPA'90, to consult the NRA if it has reason to believe that a particular landfill site is causing pollution of the water environment. Section 61 was to have been implemented in April 1993, but is now likely to take effect from 1st May 1994.

Section 143: Registers of Land which has been put to a Contaminative Use

5.16 This Section would have imposed a duty on Local Authorities to compile and maintain, in accordance with set Regulations, registers of land which is or has been put to a *contaminative* use. This is defined in the EPA as "...any use of land which may cause it to be contaminated with noxious substances". Initially registers were to be compiled as of 1 April 1992, with a completion date of 1 April 1993. However, it was announced in March 1992 that the compilation date was to be postponed to enable further consultation. A revised set of draft regulations was again distributed for consultation on 31 July 1992 (Appendix 2). The new proposals differed from the original ones in a number of ways. Firstly, the list of contaminative uses was a more limited and was intended to include only those land uses which had a high probability of causing contamination, although all such land would be placed on the register. The content of the registers would have included only the past and present uses of the land and any information on previous investigation or treatment. There was to be no appeal system to have the entry removed. Much disquiet was expressed during the first consultation process about the fact that there was no such provision, because it was felt that there would then be no incentive to undertake remedial work. The new proposals continued along this line, but made some allowance for sites which had undergone remedial treatment to be separately recorded. Thus the new proposals introduced a two part register: Part A would have recorded land which had neither been investigated nor treated, and Part B would have recorded land which had been investigated and/or treated for any contamination found. Part B was designed to provide prospective buyers and any other interested parties with access to additional information. The *caveat emptor* principle would still have applied and it was very unlikely that a provision to remove a property from the register would have been made.

5.17 The methodology which was to be used for the compilation of the register resulted from a pilot study which was carried out on behalf of Cheshire County Council. The County Council had approached the DoE in 1986 with a number of concerns about contaminated land in the area. It was agreed that the area (covering approximately two and a half local authority districts) would be used in a pilot study to identify a methodology for the Section 143 registers. The study was funded by a DoE grant and undertaken by a group of consultants. It was desk-top based and used historical maps and local information to compile knowledge on potentially contaminated land uses in the area. It identified 1500 sites and although the data from the study have never been published, it is available to the public for the purposes of land development and is also used by Cheshire District Councils to determine landfill gassing sites.

5.18 The NRA generally welcomed the proposed introduction of registers, because it would assist in its role as consultee, and for the more positive approach which was to be taken generally to the issues of contaminated land. It noted, nevertheless, that the scope of the Register had been significantly reduced and its precise nature had to be recognised for what it was because, notwithstanding the proposed Parts A and B,

- the Register would not contain information relating to land which was known to be contaminated, and yet

- would contain information relating to land which may well not have been contaminated.

The NRA also commented specifically on the exclusion of certain uses of land which are known to give rise to water pollution problems and included some of the following:

(i) sites of the manufacture of refining metals other than lead and steel - copper refineries especially are a significant source of residual pollution in the Midlands (Cu, Ni and Zn), and it is anticipated that other refinery sites present a similar hazard;

(ii) sites where there has been widespread use of chlorinated solvents, such as for car manufacture, the metal finishing industry, tanneries, dry cleaning processors etc. - groundwater pollution in urban areas largely relates to solvent pollution and the proposals ignored this issue; and

(iii) military establishments - a number of large military establishments located on aquifers have significant groundwater pollution problems from solvents and hydrocarbons, as a result of spillage on to the land.

5.19 The NRA also noted that it would be preferable for an assessment of present conditions of contamination to be held on a register and that developers should undertake an environmental impact assessment (EIA) on the current status of contamination and the potential impact on controlled waters. Such comments were of a similar nature to those made by the Royal Institute of Chartered Surveyors, who proposed land quality statements which would be attached to any planning application.

On 24 March 1993 the government announced that the proposals for Registers were being withdrawn and that it would conduct a wide ranging view of the arrangements for dealing with contaminated land and related liabilities.

The Planning System

5.20 The UK has always taken pride in its Town and Country Planning system. The majority of this was introduced in the post war years and aims to control the development and use of land in the public interest. The system is a complex one and is concerned not only with conventional commercial and residential development, but also broader land use issues such as mining, waste disposal, the provision of public space, amenity, and so on.

5.21 The planning system is operated on a number of tiers, ranging from national policy and guidance, at the top, to local plans detailing the Local Authority's specific proposals for the development and use of land. The system in England is based on *Development Plans*, consisting of *Structure Plans* which set out a broad framework at the county level, and *Local Plans* which provide a more detailed framework at the district level. Each county must also produce separately, a waste local plan and a minerals local plan, although the two types of local plan may be combined. These are to inform on county decisions for those matters on which the county is the development control authority. In Wales the position is similar, but the Districts are responsible for decisions on waste matters. In metropolitan areas the elements are combined together to form Unitary Development Plans (UDPs). Essentially these are merely a variation in nomenclature and all are governed by common national and regional policy. These complexities are explained in Planning Policy Guidance No.12, "Development Plans and Regional Planning Guidance" (DoE, PPG12, 1992).

The Planning System and the NRA

5.22 Certain parts of the planning legislation require Local Planning Authorities in turn to consult the NRA. The NRA is a statutory consultee in the preparation of development plans under the Town and Country Planning (Development Plan) Regulations 1992. It is also a statutory consultee for certain types of planning application under the Town and Country Planning General Development Order 1988, Amendment No.2 (SI 1989/1590). In other parts of the system, the NRA may be consulted if it is considered that there could be implications for the water environment in a particular development proposal. In such cases advice given by the NRA may be taken into account when it is considered that water issues are material considerations to the planning application. The NRA in any case is very keen to work closely with Local Authorities on planning issues relevant to NRA functions, regardless of the consultee status.

5.23 Table 5 illustrates the main areas within the planning system where consultation is statutory and those where it is not, but where NRA interests should be considered. The NRA is of course also involved with numerous individual planning applications - some of which are relevant, some of which are not.

TABLE 5 - NRA CONSULTEE STATUS IN THE PLANNING SYSTEM

STATUTORY CONSULTEE STATUS	NON-STATUTORY STATUS
• DEVELOPMENT PLANS: - Structure Plans - Local Plans - Unitary Development Plans - Waste Local Plans - Mineral Local Plans • GENERAL DEVELOPMENT ORDER 1988 • CERTAIN PLANNING APPLICATIONS	- Redevelopment of Contaminated Land - Recommended Consultee, DoE Circular 22/88 (Appendix C)

The Planning System and Contaminated Land - How it Works

5.24 The planning system has a vital role to play in the overall mangement of contaminated land and has been one of the main tools used in remediation of contaminated land sites in the UK. Remediation has been market led and the clean-up has been very much dictated by both the demand for redeveloping land and the end use proposals for that land. Many of the sites over which the planning system has ultimate control are those which can be defined as derelict and over which the NRA has no direct legal powers. Derelict sites may also contain contaminants which in their undisturbed state present no pollution problem but which, during redevelopment, may become mobilised and subsequently create widespread pollution. The NRA is aware that a precautionary approach to pollution control in planning decisions is of great importance and that the planning system itself is a strong precautionary mechanism, possibly as effective in the longer term as the relevant pollution control system itself.

5.25 General guidance on the redevelopment of contaminated land sites is given to local planning authorities and developers in a number of notes published by the Inter-Departmental Committee on the Redevelopment of Contaminated Land (ICRCL). These notes range from the general, such as ICRCL 59/83 on "The assessment and redevelopment of contaminated land", to the specific, such as those relating to landfill sites, gasworks, sewage works and farms. However, for NRA purposes, these notes make little reference to the water environment and without exception give no active advice, other than that one should inform the relevant body if it is felt there might be a risk to the water environment. Policy advice is also given to Local Authorities in a number of DoE circulars covering matters such as the redevelopment of contaminated land (DoE Circular 17/89), and the use of planning conditions in granting planning permission to sites which are found to be contaminated.

5.26 Development plans may also specifically include policies for the reclamation and redevelopment of contaminated land. In exercising their discretion to refuse or grant a planning application, Local Authorities may require a site's history, and a detailed investigation of the site, to provide adequate information for the proposal to be determined. Given that there is reason to request a remedial programme, the Local Authority may then require action to do so, under the planning laws, and may attach conditions for remediation to the granting of planning permission.

5.27 The NRA is a statutory consultee for some decisions on planning applications and will recommend and advise on appropriate conditions when consulted, but the practical application of this arrangement has been widely criticised for a number of reasons. Firstly, as was pointed out by the House of Commons Select Committee in their report "Contaminated Land"(1990), there is in general a severe lack of information held on contaminated land in the UK. As a result, if site history is not available or if a site investigation has not been requested, it is difficult for bodies such as the NRA to make sensible and appropriate recommendations. In such cases the need for consultation with the NRA may therefore not be appreciated or the NRA may have insufficient

FIGURE 1: THE ABOVE GROUND VIEW OF THE OLD LANDFILL SITE.

FIGURE 2: REMOVAL OF CYANIDE DURING REMEDIAL WORKS.

FIGURE 3: ABOVE GROUND VIEW OF THE OLD WASTE TIP.

FIGURE 4: DISCHARGE POINT TO LOCAL WATERCOURSE NOW CONTAMINATED WITH PCBS.

FIGURE 5: NEW INDUSTRIAL ESTATE ON OLD CONTAMINATED LAND.

FIGURE 6: UNVEGETATED LAND ON DERELICT EXTRACTION AND REFINING SITE FOR HEAVY METALS.

FIGURE 7: SURFACE WATER RUN-OFF WHICH HAS THE POTENTIAL TO POLLUTE THE RIVER TAMAR.

FIGURE 8: REFINERY SITE AND ADJACENT FOREST PLANTATION.

FIGURE 9: BLYTH HARBOUR, WHERE HIGH LEVELS OF PCBS HAVE CAUSED EXTENSIVE POLLUTION.

FIGURE 10: CONTAINING THE POLLUTION AT BLYTH HARBOUR.

FIGURE 11: SMELTING WASTE BRIQUETTES - USED TO RAISE BANKS ON PARTS OF THE SITE.

FIGURE 12: SURFACE WATER DISCHARGES - EN ROUTE TO SEVERN ESTUARY.

FIGURE 14: SEEPAGE OF CONTAMINATED WATER INTO RIVER TAME.

FIGURE 13: MAIN DRAIN - SEEPAGE TO RIVER TAME.

FIGURE 15: DAFFODIL GROWING AREAS, NOW CONTAMINATED WITH BIOCIDES.

information to support sound comments. More importantly, any risk of water pollution will not be detected and a potential problem will not be managed. It is for this reason that the NRA had generally welcomed the setting up of a register under the EPA'90. Enforcement powers have not been used sufficiently, and have also been criticised as being complex or inadequate, and although recommendations submitted by external bodies such as the NRA may be included in a planning permission, there is still no guarantee that these works will be carried out. With the number of contractors and sub-contractors that are often involved in a site redevelopment, it is not an easy task to ensure compliance with conditions. This further adds to the complexity of the situation. However, the Planning and Compensation Act 1991 did introduce some new enforcement powers and PPG 18 sets out guidance on the enforcement of planning controls (DoE, PPG 18).

5.28 Because of the general dissatisfaction with these aspects of the planning system, other options have been used to reinforce standard planning conditions. Although still part of planning legislation, the use of Section 106 Planning Agreements under the Town and Country Planning Act 1990 has been encouraged by many independent bodies, including the NRA. The use of a planning agreement may help in overcoming planning objections to a development proposal and may enable a permission to be granted which would otherwise be refused. Furthermore, a planning agreement is a legal agreement, not only with the party who enters into the agreement, but an agreement which runs with the land itself. Therefore in the event of the disposal of the land to another party, the new owner buys the liability attached to the land as well. A planning agreement may include a clause requiring a financial bond from the developer as security in the event of default from the conditions of the agreement. However, it should be noted that section 106 obligations should only cover what is needed to enable a development to go ahead. There should be a direct relationship between the requirement and the development: section 106 obligations should not be used to request unnecessary benefits.

Recent guidance on the Planning System

5.29 A Planning Policy Guidance Note "Planning and Pollution Control" is due to be issued soon and will take a comprehensive view of all aspects of the planning system and how environmental and pollution control issues should be incorporated into everyday planning decisions. The NRA received the draft document in a very positive light and commented on it extensively. In areas where the NRA does not have statutory consultee status, such as contaminated land, the NRA recommended that consultation be given nonetheless, in the best interests of planning and development.

6. PAYING FOR CLEAN-UP

6.1 Some of the issues which have been driving and moulding contaminated land policy in the UK relate to the economic implications which clean-up will bring. The general objections to the DoE's original proposals for Section 143 registers illustrated these concerns very clearly. Similarly, the complications which will arise from incorporating personal liability are not only causing concern, but as has been commented on widely, it is unlikely to cover the extent of the problem which needs to be solved. Additionally there is the argument that because of the widespread historic nature of the problem, industrialists from the earlier part of this century cannot be held liable for what may have been common and acceptable practices then, and are not today. Hence the problem of who pays the bill is almost as daunting as that of identifying and prioritising the number of sites in the country.

6.2 In some situations public funding is available for projects indirectly related to contaminated land. One of the main criticisms of these funds is that they are targeted at specific projects (eg. regeneration of derelict urban areas) and that they are available mostly in the form of grants, which are only payable on completion of a project. Capital investment must therefore be available prior to the commencement of a project; for many, this is simply not possible. And, because the schemes are only indirectly linked to contaminated land, the criteria for grants will not always include contamination and its remediation. Therefore what is available currently in the way of public funding is limited with regard to the scale of the contaminated land problem, and even more so in relation to water pollution arising from contaminated land sources. Private funding, and that via the financial system, is critically dependent upon clear government statements on what is required and the criteria to be used to determine whether or not any work has been successfully completed.

Derelict Land Grant (DLG)

6.3 Government policy on derelict land is to return such land to a beneficial use as soon as possible (DoE,1991). The Derelict Land Grant (DLG) programme is one of the sources of finance available to achieve this objective. It is a Government fund payable by the Secretary of State for the Environment in England to Local Authorities and other bodies, such as private companies and voluntary sector organisations, for the reclamation of land throughout England. A similar system operates in Wales, funded by the Welsh Office and administered by the Welsh Development Agency. In April 1991 the Government decided that the priorities of the DLG programme should be revised to allow for greater flexibility in project selection within the context of locally developed reclamation strategies. These new priorities will allow for a more substantial level of the grant to be made available for projects aimed at improving the environment; however, such projects would be included as part of overall reclamation programmes. Schemes for other environmental purposes, such as those designed to improve facilities for public relaxation or recreation, or aimed at nature conservation, will also be supported.

6.4 Sites which have been contaminated by industrial or other development may need to be tackled urgently in some circumstances, but treatments to destroy, remove or contain contaminants can be very expensive. The DLG programme already deals with many contaminated sites, but the Government has accepted that there is a need to make faster progress in dealing with contamination, subject to the constraint of the level of resources available to the DLG programme as a whole (DoE, 1991).

6.5 Similarly, it has now been recognised that there is a need to utilise some of the DLG to carry out remedial works to deal with the effects of landfill gas arising from closed landfill sites. Such works will be eligible for DLG where they form part of a scheme to reclaim derelict land for a beneficial use (DoE, 1991). In most cases remedial works will be carried out on the land which is to be reclaimed, but consideration will also be given to adjacent land (which may itself not be derelict) where it is necessitated by the effects of the reclamation works; for example, the provision of a barrier to prevent gassing on the reclaimed land may cause gas to migrate to an adjacent site, necessitating additional remedial works (DoE, 1991).

6.6 Priority within the DLG programme will be given to land which in its present condition reduces the attractiveness of an area as a place in which to live, work, invest or, because of contamination or other reasons, is a threat to public health and safety or to the natural environment. Such land should be capable of being used to provide for development, amenity value for the community, or to contribute towards nature or history conservation (DoE, 1991).

6.7 Central resources available in England 1992/93 for DLG were £106 million, an increase of 21% on funds available for the previous year. Similar funds are provided by the Welsh Development Agency (WDA) in Wales and amounted to £32 million in 1992/93. Additional funding is available via the Urban Development Corporations, City Grant, and the Urban Programme, giving a total figure of approximately £150 to £200 million per annum.

Operational priorities and objectives

6.8 The Government has set out an operational procedure to be followed as part of an area's overall strategy on land reclamation. This requires the Local Authorities in England to develop a strategic approach to reclamation involving both private and voluntary sectors. In urban and urban fringe areas, strategies should place emphasis on both economic regeneration and environmental improvement. Reclamation for amenity use or environmental improvement can also be supported, where such use is designed to enhance the attractiveness of the area for investment or as a place in which to live or work. Rural areas, particularly areas of high scenic quality, will be supported as will schemes which will foster development. In carrying out any scheme, a programme must secure maximum effectiveness, efficiency and economy in relation to public expenditure (DoE, 1991).

6.9 These changes in Government policy now provide some scope for the NRA to have an indirect input into reclamation programmes, and will for the first time enable it to benefit from some of this financing. It would be difficult for the NRA to apply for a grant on an individual project for the water environment; it would have to work as part of an overall programme of reclamation and improvement. This, of course, is not impossible to co-ordinate, because the NRA works closely with Local Authorities on many other aspects on environmental management. But it does present some problems. For example, dealing with what may well be an urgent priority as far as protecting groundwater is concerned may not be of any urgency above the ground. Similarly, scarred land from an industrial process may well blight an area visually, without having any impact on the water environment. Co-ordinating schemes of equal priority to all interested parties may therefore be difficult to achieve and, as the underlying message of the operational priorities seems to be that of producing a multifunctional end product which will attract private investment, NRA proposals may not always match operational objectives. This situation could be considerably improved if the guidance was further revised such that DLG should be spent if it was required to achieve a statutory Water Quality Objective (WQO), derived through a Catchment Management Plan (CMP), because the existence of such contaminated land, if known, would have been taken into account when the WQO was statutorily set.

How DLG currently works

6.10 In England, central Government grants are available under Section 1 of the Derelict Land Act 1982 to Local Authorities and other public bodies, voluntary organisations, private firms, and individuals for the reclamation of derelict land. Grant is paid at the appropriate rate on any *net loss* incurred in carrying out an approved reclamation scheme.

6.11 According to the Department of the Environment's publication "DLG - How it Works" (HMSO, May 1991) the term *net loss* implies that the grant is only payable on schemes which incur a loss after taking account of the post reclamation value of the land; that is, in cases where the cost of reclamation exceeds the enhancement in land value. The net loss will be determined by off-setting the approved total expenditure by the increased value of the land attributable to reclamation. Grant will not be paid towards any expenditure in excess of the approved figure. In practice, however, the grant is paid to local authorities on a gross basis and the increase in value of the land attributable to reclamation is recovered as "after value" when the reclaimed land is disposed of or otherwise brought into use. In the case of non-local authority applicants, the increase is calculated before the grant is approved and deducted from the grant paid. An example of how the grant is assessed is given in Table 6 using the 50% grant basis.

6.12 Site value is based on open market prices; the value of the land before reclamation is assessed on the basis that grant will not be payable. Applications are considered having regard to local needs. Account will also be taken of the need for the grant, the value for money, and the priority of the grant in relation to other schemes and funds available for that financial year.

6.13 Having examined the proposals and procedures which make up the Derelict Land Grant programme, it is obvious that although the new scheme of proposals is an improvement and at least recognises the inclusion of environmentally orientated projects, there are still several shortfalls as far as NRA interests are concerned. There may be great difficulties in finding projects which will be of priority to the NRA and all other interested parties. Secondly, as the grant is only paid on net loss incurred, it is likely that most parties engaging in clean-up of derelict land will actually own the land; and reclamation costs may be off-set against the sale of the improved property. The NRA does not have the finances to purchase land with the sole intention of cleaning it up and then selling it again - nor is this a NRA function. An additional problem lies in the fact that much of the emphasis of the DLG is to get land back to a beneficial and marketable state as soon as possible. In the case of groundwater pollution in particular, remedial works may take a substantial time to take effect, perhaps up to ten years in some cases. Finally, the DLG focuses on derelict land; contaminated land from derelict sites is only part of the problem as far as risks to the water environment are concerned.

TABLE 6 - DLG - HOW IT WORKS

Source: DoE, 1985.

A.	Cost of reclamation works £500,000
B.	Value of site before works £650,000
C.	Value of land after reclamation but before development £800,000
D.	Increased value of land (C-B) £150,000
E.	Eligibility of grant (A-D) £350,000
F.	**Grant Payable (50%) of E £175,000**

City Grant

6.14 The City Grant has been specifically aimed at private sector capital investment development projects in priority areas in England which are:

- above £500,000 total project value;

- unable to proceed because costs exceed values;

- providing jobs, private housing and other benefits; and

- generally within the area designated as the "urban core".

The Welsh equivalent of City Grant is the Urban Investment Grant.

6.15 Grant aid through City Grant extends the area of assistance beyond reclamation to hard development. Projects can be industrial, housing, leisure or commercial, but in all cases must demonstrate a positive contribution to the economic development of the urban area, with the grant aid structured such that a significant proportion of private sector funding was directed to the scheme (Ironside Environmental Consultants, 1992). Hence the grant has been of no assistance to the NRA or to other organisations not in the private sector and not in the business of "development".

Urban Regeneration Agency

6.16 Responsibility for the Dereclict Land Grant programme will be taken over by English Partnerships (legally, the Urban Regeneration Agency) in April 1994. English Partnerships will also take over City Grant and the work and assets of English Estates, and the three programmes will be subsumed into a unified investment regime. English Partnerships' aim is to secure the regeneration of land which is derelict or vacant, in both urban and rural areas, and it has the powers to provide financial assistance to anyone involved in this process. It will establish its own specific priorities through an annual corporate plan to be discussed with Ministers. English Partnerships will honour all existing commitments made under the programme it is taking over.

Supplementary Credit Approvals (SCAs)

6.17 Supplementary Credit Approvals (SCAs) constitute a system which may also be used to clean up contaminated land by means of public funding. SCAs are credit approvals which are made to Local Authorities to borrow for the specific purpose for which approval is given over and above any borrowing permitted under their basic credit approvals. The SCA programme does not apply in Wales; instead local authorities are able to fund waste projects from their general credit approvals, if they consider them to be a priority. The SCA programme, which includes provision for contaminated land, comes under the SCA Waste Programme 1992/93. A budget of £100 million has been set aside for the programme and from this £25 million has been allocated to the category of Landfill Gas/ Contaminated Land. How the sum will be divided between the two parts of the category will depend on bids. It is noteworthy that in the autumn of 1990, SCAs totalling £28 million were set aside for landfill gas alone, and by June of 1991 bids of £23.5 million had been received from 68 Local Authorities. The proportion of the funding currently available, which will go to remedial programmes for contaminated land, will therefore depend very much on how serious Local Authorities perceive their contaminated land problem relative to their problem with landfill gas. Because the scheme is specific to Local Authorities, and their equivalents in Wales, the NRA has no means of benefiting from it. The NRA nevertheless welcomes the programme and the recognition that contaminated land is getting.

7. THE NRA - NATIONAL AND INTERNATIONAL RESPONSIBILITIES

7.1 Spending money to clean up contaminated land will to some extent be dictated by both national and international responsibilities: these are likely to determine priorities. As part of the overall mechanism to safeguard the water environment as a resource in England and Wales, the NRA's cornerstone for its water quality responsibilities under the WRA'91 is that of the setting and achievement of WQOs which, for the first time, will have a statutory basis and will be the means by which a national plan of improvement will be implemented. Under the WRA'91 it is for the Secretary of State to prescribe (by Regulations) a system of classifying the quality of controlled waters, and then to set that classification for the purpose of maintaining and improving the quality of the waters to which the particular objective relates. The Secretary of State indicated that he wished to establish the first stage of this programme (the classification scheme) in 1993 with a rolling programme thereafter for setting objectives, starting with rivers. The scheme is now likely to be introduced during 1994. Assessing the impact on water quality of leaching from contaminated land is therefore a necessary precursor to making recommendations for the setting of WQOs; thus the legal and practical ability of the NRA to deal with the problems for such waters caused by contaminated land will be a constraint which could dictate the level at which WQOs could realistically be set in some catchment areas.

7.2 The means and purposes of setting WQOs are covered by Sections 82,83,84 and 213 of the WRA'91. Collectively they provide the means by which a scheme of setting different types of water quality objectives, for different reasons, by different dates, and incorporating EC legislation, can be developed and used.

7.3 An outline scheme of such WQOs has been published by the Department of the Environment and Welsh Office for public consultation (DoE, 1992), based on the recommendations of the NRA to the DoE/WO following its own public consultation document (NRA, 1991). The essential features are that the quality objectives of individual stretches of water should be based on the various uses to which water is or could be put. A means of generally assessing the quality of waters on the basis of their physical, chemical and biological characteristics, at five-yearly intervals, is also discussed. The NRA has agreed to recommend stretches of water for statutory WQO setting via sets of Catchment Management Plans (CMPs). Thus the CMPs effectively constitute the first step in their derivation and provide the means for examining all of the factors - discharges, abstractions, and land-use management - which could affect water quality. Thus areas of contaminated land within catchments will be assessed relative to other potentially polluting sources, and within an appraisal of the benefits of improving water quality and the effort required overall to do so. The first 'use' Water Quality Objectives to be set, within a limited number of catchments, will be those related to protection of the Fisheries Ecosystem.

International Responsibilities

7.4 As a member of the European Community, the UK is legally bound to implement and comply with all relevant EC legislation. The Treaty of Rome (1957) did not originally contain any provisions concerning the environment, but in 1986 it was amended by the Single European Act which inserted specific provisions (Section 130) on which environmental legislation could be based. The Treaty was further amended by the Maastricht agreement, one of the most important articles being Article 130r, which reiterated the objectives of Community environmental policy set out by the following principles:

- the precautionary principle;

- preventive action; and

- the polluter pays principle.

These three "principles" cover all environmental legislation to be incorporated into national legislation arising from EC Directives.

7.5 There are a number of EC Directives with which the NRA must seek to achieve compliance in England and Wales, which it does as the 'competent authority'. Table 7 lists the Directives and proposed Directives which are most relevant to water pollution which may relate to contaminated land sources, plus a list of other international agreements.

TABLE 7 - CURRENT AND DRAFT EC DIRECTIVES RELATED TO WATER POLLUTION FROM CONTAMINATED LAND, AND OTHER INTERNATIONAL AGREEMENTS

GROUNDWATER RESOURCES	SURFACE WATER RESOURCES
Council Directive 80/68/EEC - Groundwater Protection.	Council Directive 76/464/EEC Dangerous Substances.
Draft Council Directive COM (91) 102 and COM(93) 275- Landfill of Waste.	
Draft Council Directive COM (91) 219 - Civil Liability for Damage Caused by Waste.	- North Sea Conferences (Red List Substances) - Paris Commission 1987.

Groundwater

7.6 As an acknowledgment of the increasing value of groundwater as a resource and the immense problems associated with not only identifying contamination but also treating it, the EC introduced in December 1979 a specific Directive (80/68/EEC) for the protection of groundwater from pollution caused by dangerous substances; a date of December 1981 was set for compliance. The requirements of the Directive are to avert the pollution of groundwater by providing for the prevention of the discharge of List I substances and the limitation of discharges of substances in List II (Table 8). The method of control for List I substances is to prohibit their discharge directly into groundwater and, if necessary, prevent or regulate via authorisations any disposal of the substances or other operation that might lead to them entering the groundwater after percolation through *the ground or subsoil* (known as indirect discharge) so as to prevent such a discharge. Control for List II substances is by means of investigating all direct discharges and disposal or other operations which may lead to indirect disposal, and to regulate such activities via authorisations.

7.7 The Directive is unusual in that it does not set specific standards for concentrations of substances in discharges or the receiving environment. In addition, it provides for substances from the List I families and Groups to be treated (exceptionally) as List II substances on the basis of low risk of toxicity, persistence and bioaccumulation. The Directive has been criticised for a lack of technical detail and the absence of criteria for distinguishing between List I and List II substances. The lack of technical information has made it difficult to decide on specifications for monitoring frequency and methods of testing. In England and Wales the limiting of inputs of List II substances is governed by the NRA's consent system.

7.8 In October 1990, following correspondence with the European Commission over implementation of the Directive, the DoE and WO issued a consultation paper proposing a National Classification Scheme for dangerous substances under the Directive. The Departments also issued further guidance in DoE Circular 20/90 (WO Circular 34/90) on implementation of the Directive including a revised basic procedure for deciding whether a substance should be allocated to List I or List II. The Circular did not however include precise rules for classifying substances. These were laid down in a direction issued in July 1992, under the WRA'91, by the DoE and provided criteria for the NRA to classify dangerous substances for the purposes of the Directive. It was also confirmed that the NRA would administer the National Classification Scheme. The NRA has convened a National Advisory Group to examine relevant data for individual substances within the families and groups which make up List I. The National Advisory Group will then advise the NRA on: (a) which individual substances may be placed on List II due to low toxicity, persistence and bioaccumulation and; (b) which should rightly remain on List I. The classifying process will be an on-going exercise.

TABLE 8 - LIST I AND LIST II - DANGEROUS SUBSTANCES

Source: Council Directive 80/68/EEC

LIST I SUBSTANCES (PROHIBITED)	LIST II SUBSTANCES (LIMITED)
- Organohalogen compounds and substances which may form such compounds in the aquatic environment. - Organophosphorus compounds. - Organotin compounds. - Substances which possess carcinogenic, mutagenic, or teratogenic properties in or via the aquatic environment. - Mercury and its compounds. - Cadmium and its compounds. - Minerals oils and hydrocarbons.	- The following metalloids and their compounds:- Zinc, Copper, Nickel, Chromium, Lead, Selenium, Arsenic, Antimony, Molybdenum, Titanium, Tin, Barium, Beryllium, Boron, Uranium, Vanadium, Cobalt, Thallium, Tellurium, Silver. - Biocides and their derivatives not appearing in List I. - Toxic or persistent organic compounds of silicon. - Inorganic compounds of phosphorus and elemental phosphorus. - Fluorides, cyanides. - Ammonia and Nitrites. - Substances which have deleterious affect on the odour or taste of groundwater.

7.9 The Groundwater Directive was criticised in the EC Fifth Action Programme on the Environment (1991) which stated that it "is not achieving its objectives and these [water] resources are under growing threat from over exploitation and pollution". The EC Environment Minister's seminar on groundwater held in November 1991 produced a declaration setting out an Action Plan for groundwater protection. This declaration was affirmed by a Council of Ministers Resolution in February 1992. The Declaration sets out a number of objectives, those relating specifically to the contamination of groundwater being as follows:

- preserving the quality of uncontaminated groundwater and preventing the further deterioration of contaminated groundwater;

- restoring contaminated groundwater and soil to a quality required for drinking water where appropriate; and that

- water management must be integrated with other environmental policies dealing with matters such as agriculture, industry and tourism. (This is undoubtedly also focusing on land use planning issues and therefore on contaminated land policies, be they for the redevelopment of existing sites or for the clean-up of operational industries).

7.10 The Ministers asked the Commission to prepare detailed proposals to enable the Action Programme to be implemented at Community and national level by the year 2000. The proposals need to include a revision of the current groundwater Directive and the incorporation of groundwater protection provisions in other relevant Directives.

7.11 Given that this level of commitment is being exercised by the EC, it will be the case that clean-up and further prevention of pollution to groundwater from contaminated land sites will be something that will have to be approached proactively and in addition to the national standards set by the implementation of WQOs. The NRA has also introduced its own Groundwater Protection Policy as a proactive preventative measure (NRA, 1992), the use of which is discussed in the next chapter.

Draft Landfill Directive - COM(91) Final

7.12 In addition to having its own Directive, groundwater should shortly be further protected from one source of pollution by the introduction of a Landfill Directive. This is currently in draft form as COM (91) Final - Regulating the Landfill of Waste. The main aim of the Directive is the eventual harmonisation in all Member States of technical and environmental standards for landfill and to ensure a high level of protection for the environment in general, and for the protection of soil and groundwater in particular (NSCA, 1992).

7.13 This Directive will be of particular importance to the NRA; many of the pollution problems from contaminated land sites result from that section of contaminated land classified as landfill. Of 186 sites identified as almost certainly actively polluting (with a few suspected of polluting) at present in the NRA's Severn Trent Region alone, 45% were classified under the landfill category.

7.14 The draft Directive classifies waste according to origin (municipal or industrial) and characteristics (hazardous, non-hazardous and inert). Landfill sites are also classified, such as those for hazardous wastes, for municipal wastes and so on. The concept of "mono-fill" (ie only wastes of a comparable composition could be deposited) and "multi-disposal" are introduced. Additionally, all criteria for the closure, subsequent monitoring, and financial liabilities are included.

Draft Directive on Civil Liability in Waste COM(91)219

7.15 Mirroring the recent accelerations in interest in contaminated land in the UK, the EC has also proposed a Directive on Civil Liability for Damage caused by Waste. A text of the Draft Directive was published in 1989, and an amended version - COM (91)219 - in 1991. The main thrust of the Directive is to impose civil liability irrespective of fault on the producer of waste which causes damage to persons or property or impairment of the environment (defined as any significant deterioration).

The actual producer may escape liability if it can be shown that the damage occurred after waste had been transferred to a licensed disposal site.

The waste producer would also remain liable for any damage to the environment if the waste disposal facility can prove that it was given incorrect or false information about the waste. The Draft Directive also provides for environmental protection and interest groups to take legal action to ensure that any lasting damage to the environment is remedied; it will be left to individual Member States to introduce the legislation which would implement this right. There has been no progress on this draft directive for several years. This proposed Directive has several other implications regarding insurance and what type of cover insurance companies are prepared to give industry against civil liability. More recently the EC has produced a Green Paper on Remedying Environmental Damage which may well subsume this Directive. The paper has two main objectives:

a) to examine the issue of civil liability as a legal and financial tool to be used to rectify and prevent damage caused to the environment, a concept which ties in with commitments made in the 4th and 5th Environmental Action Programmes; and

b) to examine the possibilities of remedying environmental damage not met by civil liability.

It also discusses the difficulties of applying liability and in defining what environmental damage is and who or what is causing it.

Surface Waters

7.16 Although part of the same system and resource, both the impact of pollution and the legislation to protect surface waters are different from those of groundwater resources, but because it is usually surface waters which highlight the more easily identifed pollution problems associated with contaminated land, the degree of monitoring carried out on rivers and water courses tends to be greater than monitoring of groundwater.

Dangerous Substances Directive (76/464/EEC)

7.17 Discharges to surface waters and other waters are covered by the "Framework" Directive (76/464/EEC), and subsequent "Daughter" Directives on pollution caused by certain dangerous substances discharged into the aquatic environment. This Directive does not apply to groundwater, and although it also contains List I and List II substances, these should not be confused with those relevant to the Groundwater Directive, despite some overlap. The "Framework" Directive sets out List I substances, often referred to as "Black List" substances. These are considered to be so toxic, persistent, or bio-accumulative that all necessary action should be taken to eliminate them from entry to the aquatic environment. Elimination is generally achieved through "limit values" set in daughter Directives. Daughter Directives have been introduced for cadmium, mercury, hexachlorocyclohexane (HCH), all the "drins" group of pesticides (aldrin, endrin, dieldrin and isodrin), and DDTs, carbon tetrachloride, pentachlorophenol, HCB, HCBD, chloroform, 1-2 dichloroethane, trichloroethylene, perchloroethylene and trichlorobenzene. Not all the substances in List I are banned, many are "controlled" by standards set out in the Directive. For List II substances the standards are set by each Member State as with the Groundwater Directive - according to national requirements. However, there is an obligation on each Member State to illustrate that the standards set represent a progressive programme of reduction of the use and presence of the substances. The procedure for setting standards may take two approaches:(i) limit values, which set an absolute standard in terms of concentration discharged; and (ii) the Environmental Quality Objective approach, whereby standards are set according to the environmental receiving capacity of the particular stretch of water. Therefore all discharges can be set at different levels, as long as the environmental water quality objective is achieved. This is the system which is used in the UK, implemented in England and Wales by the NRA's consents system. The UK is the only country in the EC which has adopted this approach or provided the EC with a reduction programme. Again, as with the Groundwater Directive, the standards set and the degree of compliance must be reported to the EC annually.

7.18 With regard to the protection of controlled waters, however, it must be borne in mind that such substances can result from any process, be it an industrial source or a leachate from a landfill site. All consented surface water discharges are the responsibility of the NRA, with the exception of prescribed processes which are authorised by Her Majesty's Inspectorate of Pollution (HMIP) under the Environmental Protection Act 1990. In such instances there is liaison with the NRA as to the conditions set in an authorisation.

7.19 The consented discharges which contain such dangerous substances are only a proportion of the problem, which can at least be accurately accounted for and controlled via the consents system. It is the diffuse sources of entry into the water environment, for example from a source such as contaminated land, which pose both the more serious threats to the water environment (because of the uncontrolled nature of discharges) and the greatest challenge to the NRA, because of the difficulties in identifying the scale of the problem, tracing the source, and combating the damage.

PARCOM and North Sea Inter-ministerial Conferences

7.20 A number of international agreements have been made in the area of pollution control and prevention which also need to be examined briefly according to their relevance to contaminated land. Dangerous and persistent substances were brought to international attention in 1972 through the London Convention and the Oslo Convention, both of which are concerned with the dumping of wastes from ships and aircraft. A further convention, the Paris Convention, was drawn up in 1974 and relates specifically to discharges of persistent substances from land based sources. The Convention identified four categories of substances relating to discharge control. The UK signed the Paris Convention in 1978. More recently, the Paris Convention and the Oslo Convention have been merged into a Convention for the Protection of the Marine Environment of the North-East Atlantic. This will require all Contracting Parties of the Paris and Oslo Conventions to adopt the Precautionary Principle, the Polluter Pays Principle and ensure that their programmes are in accordance with Best Available Technology (BAT). The Convention will also prevent these measures from causing any increase in pollution of the sea outside the area to which the Convention applies. It was agreed by Contracting Parties in September 1992, and now awaits approval by their respective Parliaments. It was put to the UK Parliament in July 1993 and is awaiting ratification.

7.21 With regard to the reduction of total inputs into coastal waters from land based sources, what has been of practical significance for the NRA is the UK's Priority Red List, first proposed in 1988. This formed the basis for action to implement that part of the 1987 Ministerial Declaration of the 2nd International Conference on the Protection of the North Sea which was concerned with the reduction of substances entering the North Sea via discharges from land. The list contained substances which occur in both EC Lists I and II, plus PCBs and several pesticides -23 substances in total. At a 3rd International Conference on the Protection of the North Sea held in the Hague in 1990, a list of 36 substances was identified for priority action; this included all those in the UK Priority Red List, with the exception of PCBs, all the current EC List I substances, 6 metals and organotin compounds from List II, plus additional pesticides and dioxins. The List was agreed by all participating members and a decision was taken to achieve a reduction of 50% of inputs of these substances to the water environment, and in some cases a 75% reduction, over the ten year period from 1985-1995. More recently, following the introduction of a system of Integrated Pollution Control under the EPA '90, a list of prescribed substances has been identified for specific control within England and Wales. This list is essentially the same as that of the UK Priority Red List except that it refers specifically to all isomers of hexachlorocyclohexane (HCH), DDT, trichlorobenzene (TCB) and to compounds of pentachlorophenol (PCP).

7.22 In view of the varying sources of these pollutants, it is clearly necessary to have a reasonable indication of the relative importance of each source such that regulation and pollution control can be used to the greatest effect. Control of discharges by limits based on concentrations of substances at specific locations in the receiving environment is not in itself sufficient; account must also be taken of the total quantities discharged because of the collective potential impact to which the environment is then committed. Hence inputs from diffuse sources such as contaminated land must be considered. The NRA is currently examining the totality of all Red List Substance inputs (Pentreath, 1992), and a full report on the subject will be produced later in 1994.

8. THE JOB TO BE DONE

Introduction

8.1 Addressing the problem of contaminated land will be a major task for the Environment Agency. This report has only briefly considered those aspects which relate to the aquatic environment, but it is likely that such aspects would feature large in any attempt to prioritise sites in need of remedial action. For surface waters, the NRA is already in the process of drawing up catchment management plans (CMPs). These plans will be used as a basis for managing all of the NRA's core functional activities at regional and area level. They will also be produced as a necessary pre-requisite for the recommendation by the NRA to the Secretary of State of specific stretches of surface water for which any Water Quality Objectives (WQOs) might be set by statute. Within such CMPs it will therefore be necessary to identify those contaminated sites which:

- are a cause of breaching existing water quality standards (WQSs);

- would be a cause of breaching any future WQSs as a result of statutory WQOs being set; or

- contribute a measurable fraction of the input of any substance targeted for reduction under existing national or international agreement.

8.2 With respect to groundwater, the NRA has already outlined its policy and practice for the protection of such water, and the need to protect potable supplies, including its policy statements specific to the problem of contaminated land. These it will continue to pursue. Thus it has already taken a number of initiatives to address the problem of contaminated land and the water environment in general, which are relevant to the "job to be done". These initiatives may be divided into two parts: the remediation of currently contaminated land sites actually causing water pollution; and the need to ensure the prevention of their contamination in the future through the use of proactive policies, standards and practices.

Cleaning up the Past

(a) Estimating the Scale of the Problem

8.3 An obvious pre-requisite to the effective and efficient management of contaminated land is the need to know the scale and size of the problem to be solved with respect to its impact on the aquatic environment within specific catchments, or above specific aquifers. This has not yet been assessed in the UK; indeed there is currently no comprehensive appraisal of how much land area in total may actually be defined and positively identified as contaminated. Nevertheless, because the number of sites affecting the water environment will only be a proportion of the total, the NRA has started a preliminary exercise to try and estimate its scale in England and Wales.

8.4 An initial study was commissioned in 1990. Given that the then ten NRA Regions varied enormously, the results of the survey - not surprisingly - were equally variable. An additional study was therefore undertaken in the Severn Trent Region to investigate in more detail the magnitude of the problem in a single area. This study produced a list of 186 sites which are known to be, or suspected of, causing pollution to surface or ground water. Another national exercise was started in July 1992, the results of which indicated that all NRA Regions could identify a number of sites which were causing water quality problems. More, however, needs to be done. Substantial amounts of information already exist. A number of surveys have been conducted on derelict land, and although derelict and contaminated land are not one and the same thing, there is likely to be a certain degree of overlap between them. Local knowledge is also very valuable. The question arises, however, as to how such data bases should be held by the future Environment Agency, bearing in mind that it now seems unlikely that any form of register will be compiled by all Local Authorities either officially or unofficially. Such data bases would need to be carefully considered, in view of the requirements of the EC Directive on the Freedom of Access to Information on the Environment (Council Directive 90/313/EEC), and to safeguard against the charge of placing a blight on the land itself. But such data bases already exist. These include data from the pilot studies in Cheshire for the DoE in 1988, which identified over 1500 sites; potentially contaminated sites in Wales (some 746) identified in a 1988 study commissioned by the Welsh Office; the results of derelict land surveys; and information already held by local authorities which was to be the basis for compiling the registers under Section 143 of the EPA'90. In the context of the aquatic environment, however, it would be useful to indicate areas specifically known to be breaching water quality standards, a record of action taken, and an indication of the success or otherwise of such action. (There is at present no requirement for the NRA to record actions taken under Section 161 of the Water Resources Act 1991, but such a requirement could be useful as a public record of positive action taken.)

(b) Estimating the Nature of the Problem

8.5 It is also important to consider the nature of the problem with respect to the potential effect of contaminated land on the aquatic environment in order to prioritise those sites which may need to be remedied, by whatever means. Apart from differentiating between those which pollute groundwater and those which may pollute surface water, the latter may be grouped between those which lead to an excessive concentration of a substance at a given point, - in other words, it contributes to a breach of a quality standard - and those which may also contribute to a substantial quantity (or load) of a substance entering surface waters over a particular period of time. Where concentration standards are being breached, the role played by contaminated land must be considered alongside point source inputs through the catchment management process. An economic assessment of necessary remediation should also be made. With regard to the totality of inputs, however, it will be necessary to wait for the NRA's current exercise on quantifying 'Red List' and other substances into coastal waters to be concluded. This exercise is intended to produce 'rank order' listings of inputs at freshwater limits and, downstream of them, from industrial and sewage outfalls, at over 350 locations. It will then be necessary to investigate further those catchments which make a significant (>1%) contribution and examine them to see if contaminated land sites are a contributory factor. Although notwithstanding such information, any site known to be a source of PCB inputs at more than trace amounts per day, because of the general evidence against this compound (Reijnders, 1986; Addison, 1989), should be noted and prioritised for further study and action. With groundwater the problem is less easily defined. Sites could at least be classified into those known to be causing significant groundwater pollution and those where pollution of an aquifer is strongly suspected. Where pollution is occurring there are then essentially three choices of action: attempt to reduce the concentration of the substance by taking remedial action; restrict the spread of the contaminant to other boreholes by maintaining (by pumping) a cone of depression; or of doing nothing. In assessing whether of not anything should be done by the NRA, it would also obviously be necessary to consider the situation as a whole with regard to securing the proper use of water resources in that area, and thus the extent to which any action would be necessary or expedient. Such a consideration would therefore include an assessment of the relative importance of the source for abstraction or augmentation of other supplies, the extent to which its loss would affect water management options as a whole in that area, the extent to which a lasting remedy could be effected, the time it would take, the cost, and the overall benefits in a water management context.

8.6 In view of the above, it would therefore be necessary for the Environment Agency to consider contaminated land with respect to those sites within a catchment which could be shown to be:

(a) polluting, or with the potential to pollute, existing groundwater sources and /or render the water contained in a significant part of a useable aquifer to be unfit for potable use;

(b) a cause of breaching a surface Water Quality Standard;

(c) in conjunction with the 'Red List reduction' exercise - a significant (>1%) contribution to the annual input of 'Red List' substances into coastal waters; or

(d) a source of more than trace quantities of PCBs to surface waters.

Other sites identified via the CMP exercise will be characterized with regard to the perceived *level of risk* which they present of contaminating water, or of preventing national or international 'targets' from being achieved. They would therefore be considered alongside other risks to water quality within a given catchment.

Such information would have to be continually revised by drawing upon data on sites suspected of causing actual pollution, or at high risk of doing so.

(c) Dealing with the Problem

8.7 Having identified and prioritized land which is sufficiently contaminated to be causing - or at great risk of causing - pollution of groundwater or surface water, the questions naturally arise as to how and by whom should the problem be tackled. With regard to the latter, in the first instance it is obviously the owner of the land who has responsibility for it. Where a company does take action to remedy a polluting site, it is also essential that it first discusses the matter extensively with the appropriate agency (currently the NRA) to ensure that further pollution does not occur during the course of the work. Where a company possesses many sites which may have been put to a contaminative use, it should also discuss the matter with - at present - the NRA in order to draw up a priority list for remediation. This will ensure that the greatest value for money is obtained from the resources available. (A number of companies have already approached the NRA in order to obtain guidance for their own desk-top exercises.)

8.8 At present if, as a result of discussions with the NRA, it appears that the owner is unwilling to undertake a plan of remedial action, then prosecution may follow if pollution is actually being caused. One of the principal reasons for carrying out a prosecution with respect to polluting offences, however, is the desire to change the behaviour of the offender, particularly where the discharge is being actively managed. Where - as may be the case with respect to some contaminated land sites - there is an ongoing pollution problem as a result of a passive input, a successful prosecution may have little effect. Indeed the scale of remedial action may greatly exceed the owners ability to carry it out. If this is the case, the next course of action should then be to consider the use of existing sources of government grant (eg DLG) to tackle the problem. This should be easier if the work had to be carried out to achieve a WQO set by statute.

The NRA will also advise local authorities on such matters in order to ensure that as DLG is spent, not only would the land benefit but the aquatic environment would benefit too - giving added value for money.

8.9 Consideration also needs to be given as to what an Environment Agency, or currently the NRA, could usefully do under existing powers. This, in turn, depends on how exactly the situation could be remedied, particularly with regard to whether it required only a *capital* spend, or if it required a long-term commitment to manage and maintain some form of treatment system and therefore required *continual current* expenditure. If capital monies only were required, then at present the NRA could:

(a) carry out the works itself under Section 161 of the WRA'91 and subsequently attempt to recover the costs from the owner;

(b) carry out the works itself under Section 161 or 162 and bear the cost itself; or

(c) if the Secretary of State empowered it to do so, under Regulations which specifically related to contaminated land, introduced under Section 92(1)(b), serve some form of improvement notice under Section 92(2) of the WRA'91.

If Section 161 powers (or Section 162) were to be used for such capital-only projects, such projects should be limited, the money being drawn from a central fund held specifically for that purpose.

8.10 Where the situation needs to be remedied by a scheme which would involve capital expenditure followed by a subsequent management and maintenance programme, the NRA has not intended to take on an operational commitment in any way. If, however, the owner was able to manage and operate the scheme but could only afford the capital investment, then such investment would only be made by the NRA on condition of the owner's commitment to manage it properly - via either a consented discharge or a relevant prohibition with conditions placed on it.

8.11 The NRA would at present only consider using its own resources (recoverable or not) in order to carry out *short-term* remedial work involving capital monies and/or *time-limited operational* activities. Such remedial work would only be undertaken if, in the case of surface waters, it had been identified via the CMP process or, in the case of groundwaters, a cost-benefit analysis study has indicated that it is in the wider interests of the NRA's resource management role to do so. The NRA would, in any case, need to obtain the approval of the DoE for any expenditure in excess of £0.5M under its Scheme of Delegation. In effect, therefore, the government itself would be in a position to adjudicate over the extent to which the NRA committed itself to long-term remedial action.

Knowing what best to do - R&D

8.12 One further consideration is that of knowing, and deciding, what the best course of action might be - such as a choice between capital schemes, or between capital and current-based schemes. Again, it is for the owner of a polluting site to provide such information; but he may not be able to do so, and the Environment Agency would in any case need to assess such schemes. It is has therefore been necessary for the NRA to carry out studies via its R&D Programme. The NRA already has established links with DoE, MAFF, and research institutes, and co-operates with projects, or sponsors, on a wide range of relevant research. In addition, the NRA has commissioned a number of research projects specific to its own interests in relation to contaminated land. Studies include pollution risk assessments, investigations into the pollution potential from contaminated land sites, remedial treatments for contaminated land, methodologies for purification of contaminated groundwater, water quality implications of waste disposal, and remediation techniques and estimated costs. A list of associated NRA R&D projects currently running or due to start, with their output objectives, is given in Appendix 3. It is nevertheless evident that more needs to be done nationally to place the problem of contaminated land within an overall framework policy based on environmental economic assessment.

8.13 It is also essential to be able to evaluate progress, and thus whether or not objectives have been met and that value for money has been obtained. Suitable and efficient monitoring programmes therefore need to be developed further, and these have a cost in themselves. The NRA has rationalised its surface water monitoring programme, and is introducing standardised statutory standards and a general Quality Assessment scheme for surface waters. A national groundwater monitoring scheme is also being developed in connection with the Groundwater Protection Policy.

Prevention in the Future

8.14 It is important to remember that contaminated land is continually being created. For the sake of the aquatic environment the NRA has attempted to ensure that such areas are minimised in the future. Cleaning-up contaminated land currently causing problems is therefore only one issue; preventing a deterioration in this situation in the future is another. Thus wherever possible the NRA has sought to be involved and influential in programmes and practices which may be part of a national preventative policy, particularly as a statutory consultee in many areas of the planning legislation, and through its own policies and initiatives.

Statutory Consultations

8.15 The NRA is currently a statutory consultee for a variety of issues and procedures which may have an impact on the water environment. Similarly it consults widely on policies and issues it is proposing which may affect other interested parties. The NRA has recently become a statutory consultee for Development Plans and Structure Plans; it is also a consultee for certain Environmental Impact Assessments (EIAs) and with regard to the granting of waste management licences under the EPA'90. Being consulted in matters which may affect the water environment provides the NRA with opportunities positively to influence land use practices and policies, for the benefit of water quality. And although the NRA remains disappointed that the recommendation to award it statutory consultee status for the redevelopment of contaminated land made by the House of Commons Environment Select Committee in 1990 has not yet been implemented, it remains keen to contribute positively on such matters whenever requested.

Regulations

8.16 Under Section 92 of the WRA'91, the Secretary of State can make regulations which are introduced in the form of a Statutory Instrument (SI) in order to prevent pollution of the water environment. Following consultations with, and advice from, the NRA concerning a large number of preventable farm pollution incidents, regulations regarding certain agricultural practices were introduced under Section 92. The regulations - The Prevention of Pollution (Silage, Slurry and Agricultural Fuel Oil) Regulations 1991 (SI324) - require certain precautions to be taken against water pollution. They require design and construction standards on agricultural silage, slurry and fuel oil stores. The regulations came into force in September 1991.

8.17 A number of pollution incidents have led the NRA to request consideration being given to the introduction of additional regulations to prevent water pollution. Such incidents have particularly involved the spillage, seepage and release of chemicals on industrial sites whilst in storage, which have resulted in pollution of both surface and groundwaters. The NRA has stated that regulations are needed to govern the industrial storage and disposal of both chemicals and their containers. It is hoped that they would cover design and construction of stores, plus bunding and containment facilities, to prevent further contamination of both soil and water. The only current regulations relating to the storage of chemicals refer to the health and safety of workers, but do not make provision for inputs into the environment. Accidents of such a nature can have serious and long-term impacts on both soil and the water environment, and it is clear that many of them could have been avoided if proper management practices were followed. The NRA will continue to seek improvements in this area of prevention and is hopeful that regulations governing these avoidable incidents may soon be introduced. Regulations under Section 92(1)(b) also need to be introduced for the specific case of contaminated land affecting water quality, essentially to allow the NRA or its successor to serve improvement notices on site owners under Section 92(2).

The Groundwater Protection Policy

8.18 In December 1992 the NRA launched its Groundwater Protection Policy; this deals with all aspects of groundwater and how it should best be utilised and protected. Dealing with contaminated land is very much part of this overall policy, which the NRA still intends to develop. The policy incorporates catchment management plans and WQOs as part of the general objectives which it aims to achieve and has a number of policy objectives specifically related to contaminated land.

8.19 As a general procedure, the NRA seeks appropriate control to protect the water environment from the redevelopment of contaminated sites through an interface with Planning Authorities under the Town and Country Planning Acts. The NRA also liaises with HMIP in respect of its powers to promote Integrated Pollution Control. The Groundwater Protection Policy essentially involves the mapping of England and Wales in order to show the extent of groundwater vulnerability to pollution, and the division of all groundwater resources into *source protection zones*. The following policies apply specifically to contaminated land issues.

(i) The NRA will encourage the implementation of effective remedial measures to prevent pollution of groundwater from any contaminated land site by virtue of existing direct or indirect discharges. Where pollution occurs, the NRA will prosecute in appropriate cases under Section 85 of the Water Resources Act 1991.

(ii) The NRA will seek to be consulted by Local Planning Authorities about any application for development, or other works, on sites likely to be contaminated.

(iii) The NRA will recommend to the Local Planning Authority that it refuses planning permission for redevelopment of contaminated land sites where water resources could be adversely affected, unless it can be demonstrated to the NRA's satisfaction that the proposals include effective measures for the protection of groundwater and surface water quality. It will also advise the Local Planning Authority where insufficient or technically weak information has been provided so that it can require the applicant to supplement them.

(iv) The NRA will seek to ensure that planning permissions contain conditions designed to protect water resources. The NRA will strongly encourage the Local Planning Authority to seek to enter into planning obligations with developers under Section 106 of the Town and Country Planning Act, 1990 (as substituted by Section 12 of the Planning and Compensation Act, 1991) to control and monitor surface and groundwater contamination during and after redevelopment of contaminated sites.

(v) The NRA will wish to ensure that any discharge, seepage, or drainage resulting from the redevelopment of a contaminated site will be of a quality and quantity such that there will be no pollution of groundwater and/or surface waters.

(vi) The NRA will review the merits and feasibility of groundwater clean-up in areas where historical industrial development is known to have caused widespread contamination, having due regard to local circumstances and available funding.

(vii) The NRA will encourage manufacturing industries and others to improve operational practices in order that unauthorised discharges to land are eliminated. Where contamination of groundwater - or of surface water through contaminated groundwater flow - is apparent, the NRA will require remedial measures to be undertaken to prevent further pollution.

(viii) The NRA will, by liaison with Planning Authorities and industry, seek to influence the location of new industrial development areas which are not vulnerable to groundwater pollution.

(ix) The NRA will, by liaison with HMIP and others, seek to ensure that authorisations granted to industry prevent future contamination of land and groundwater.

Conclusions and Summary

8.20 The impact of contaminated land has to be kept in perspective, but one of the present difficulties is knowing adequately what the correct perspective is, because the scale of the problem has yet to be fully evaluated. The following action is therefore suggested.

Assessment

- Through the NRA's catchment management planning process, the risks to the aquatic environment from contaminated land should be assessed within each Catchment Management Plan by identifying areas which are:

    - a cause of a breach of an existing Water Quality Standard, or would prevent the achievement of a statutory Water Quality Objective;

    - a significant (>1%) contribution to the annual input into coastal waters via any route of substances targeted for reduction; or

    - a source of more than trace quantities of PCBs.

- Other sites, where known, should be identified and characterised with regard to the perceived *level of risk* which they present of contaminating water or of preventing national or international 'targets' from being achieved. Where Catchment Management Plans are used to advise on the setting of statutory Water Quality Objectives, such sites would need to be considered on the same basis of benefits, and effort required to achieve such benefits, as any other source of contamination or pollution.

- Because aquifers do not necessarily match the catchments overlying them, a specific exercise is required in connection with the NRA's existing Groundwater Protection Policy. Thus the Environment Agency should build on the NRA's groundwater centre (in its Severn Trent Region) and consider the compilation of a data base specifically related to land which is known to be polluting existing groundwater sources, and of land at risk of contaminating such water. Such levels of risk should be estimated by investigative study and modelling. The programme also needs to be supported by a monitoring programme - currently being developed - related to the vulnerability of the groundwater to a deterioration in quality from any source.

Remedial Action

- Responsibility for water pollution from contaminated land sites rests with owners. Where an owner intends to remediate a site causing water pollution, he should discuss it first with the NRA. Where several sites need attention, the NRA will advise on their relative priority. If an owner of a site 'causes' or 'knowingly permits' pollution from a site, he is liable to be prosecuted.

- In cases where the owner is unable to carry out remedial works identified via the NRA's - or Environment Agency's assessment - in relation to water quality, then application should be made

to existing sources of government grant. The guidance given to Local Authorities should be
revised to ensure that such monies are allocated to sites which need to be addressed in order to
meet statutory Water Quality Objectives derived via Catchment Management Plans.

- The NRA or Environment Agency should generally only use its own resources (recoverable or
not) in order to carry out short-term remedial work involving capital monies; they should not
take on responsibility for the longer-term running of sites which would involve crossing the
barrier from 'regulator' to 'operator'. (Such a commitment would, in any case, require approval
from government for expenditure in excess of £0.5M; in which case government would itself have
considered the long-term implications of the commitment entered into.)

- Further R&D should be carried out to place the relative effect of contaminated land on the
aquatic environment, and of any remedial action required, within an overall environmental
economic assessment in England and Wales - this should be aimed at demonstrating value for
money relative to delivery of statutory WQOs, and relative to the costs of improvements arising
from the Asset Management Plan 2 expenditure of water companies. The Asset Management
Plan 2 provides a pragmatic framework on which the NRA and Water Service Companies (WSC)
can plan and cost future needs.

Other Actions

- The feasibility of Regulations under Section 92(1)(b) of the WRA'91 should be seriously
considered such that the NRA could serve some form of 'improvement notice' under Section
92(2) on owners of contaminated land.

- Regulations should also be brought in under the WRA'91, to complement those already existing
to prevent agricultural pollution at source, with respect to the storage of chemicals on industrial
and other sites where controlled waters are at risk.

APPENDIX ONE

WATER RESOURCES ACT 1991
(relevant sections)

WATER RESOURCES ACT 1991

POLLUTION OFFENCES (PRINCIPAL OFFENCES)

Section 85 - (1) A person contravenes this section if he causes or knowingly permits any poisonous, noxious or polluting matter or any solid waste matter to enter any controlled waters.

(2) A person contravenes this section if he causes or knowingly permits any matter, other than trade effluent or sewage effluent, to enter controlled waters by being discharged from a drain or sewer in contravention of a prohibition imposed under section 86 below.

(3) A person contravenes this section if he causes or knowingly permits any trade effluent to be discharged -

(a) into any controlled waters; or

(b) from land in England and Wales, through a pipe, into the sea outside the seaward limits of controlled waters

(4) A person contravenes this section if he causes or knowingly permits any trade effluent or sewage effluent to be discharged, in contravention of any prohibition imposed under section 86 below, from a building or from any fixed plant -

(a) on to or into any land; or

(b) into any waters of a lake or pond which are not inland freshwaters.

(5) A person contravenes this section if he causes or knowingly permits any matter whatever to enter any inland freshwaters so as to tend (either directly or in combination with other matter which he or another person causes or permits to enter those waters) to impeded the proper flow of the waters in a manner leading, or likely to lead, to a substantial aggravation of

(a) pollution due to other causes; or

(b) the consequences of such pollution.

(6) Subject to the following provisions of this Chapter, a person who contravenes this section or the conditions of any consent given under this Chapter for the purposes of this section shall be guilty of an offence and liable -

(a) on summary conviction, to imprisonment for a term not exceeding three months or to a fine not exceeding £20,000 or to both;

(b) on conviction on indictment, to imprisonment for a term not exceeding two years or to a fine or to both.

Section 86 - (1) For the purposes of section 85 above a discharge of any effluent or other matter is, in relation to any person, in contravention of a prohibition imposed under this section, if subject to the following provisions of the section -

    (a)   the Authority has given that person notice prohibiting him from making or as the case may be, continuing the discharge; or

    (b)   the Authority has given that person notice prohibiting him from making, or as the case may be, continuing the discharge unless specified conditions are observed, and those conditions are not observed.

(2) For the purpose of section 85 above a discharge of any effluent or other matter is also in contravention of a prohibition imposed under this section if the effluent or matter discharged -

    (a)   contains a prescribed substance or a prescribed concentration of such a substance; or

    (b)   derives from prescribed process or from a involving the use of prescribed substances or the use of such substances in the quantities which exceed the prescribed amounts.

(3) Nothing in subsection (1) above shall authorise the giving of a notice for the purposes of that subsection in respect of discharges from a vessel; and nothing in the regulations made by virtue of subsection (2) above shall require any discharge from a vessel to be treated as a discharge in contravention of a prohibition imposed under this section.

(4) A notice given for the purposes of subsection (1) above shall expire at such a time as may be specified in the notice.

(5) The time specified for the purposes of subsection (4) above shall not be before the end of the period of three months beginning with the day on which the notice is given, except in a case where the Authority is satisfied that there is an emergency which requires the prohibition in question to come into force at such time before the end of that period as may be so specified.

(6) Where, in the case of such a notice for the purposes of subsection (1) above as (but for this subsection) would expire at a time at or after the end of the said period of three months, an application is made before that time for consent under this Chapter in respect of discharge to which the notice relates, that notice shall be deemed not to expire until the result of the application becomes final -

    (a)   on the grant or withdrawal of the application;

    (b)   on the expiration, without the bringing of an appeal with respect to the decision on the application, of any period prescribed as the period within which any such appeal must be brought; or

    (c)   on the withdrawal or determination of any such appeal.

Section 92 - (1) The Secretary of State may by regulations make provision -

    (a)   for prohibiting a person from having custody or control of any poisonous, noxious or polluting matter unless prescribed precautions and other steps have been carried out or taken for the purpose of preventing or controlling the entry of the matter into any controlled waters;

(b) for requiring a person who already has custody or control of, or makes use of, and such matter to carry out such works for that purpose and to take such precautions and other steps for that purpose as may be prescribed.

(2) Without prejudice to the generality of the power conferred by subsection (1) above, regulations under that subsection may -

(a) confer power on the Authority -

(i) to determine for the purposes of the regulations the circumstances in which a person is required to carry out works or to take any precautions or other steps; and

(ii) by notice to that person, to impose the requirement and to specify or describe the works, precautions or other steps which that person is required to carry out or take;

Section 161 - (1) Subject to subsection (2) below, where it appears to the Authority that any poisonous, noxious or polluting matter or any solid waste matter is likely to enter, or to be or to have been present in, any controlled waters, the Authority shall be entitled to carry out the following works and operations, that is to say -

(a) in a case where the matter appears likely to enter any controlled waters, works and operations for the purpose of preventing it from doing so; or

(b) in a case where the matter appears to be or have been present in any controlled waters, works and operations for the purpose -

(i) of removing or disposing of the matter;

(ii) of remedying or mitigating any pollution caused by its presence in the waters; or

(iii) so far as it is reasonably practicable to do so, of restoring waters, including any flora and fauna dependent on the aquatic environment of the waters, to their state immediately before the matter became present in the waters.

(2) Nothing in subsection (1) above shall entitle the Authority to impede or prevent the making of any discharge in pursuance of a consent under Chapter II of Part III of the this Act.

(3) Where the Authority carries out any such works or operations as are mentioned in subsection (1) above, it shall, subject to subsection (4) below, be entitled to recover the expenses reasonably incurred in doing so from any person who, as the case may be -

(a) caused or knowingly permitted the matter in question to be present at the place from which it was likely, in the opinion of the Authority, to enter into any controlled waters; or

(b) caused or knowingly permitted the matter in question to be present in any controlled waters.

(4) No such expenses shall be recoverable from a person for any works or operations in respect of water from abandoned mines which that person permitted to reach such a place as is mentioned in subsection (3) above or to enter any controlled waters.

(5) Nothing in this section -

(a) derogates from any right of action or other remedy (whether civil or criminal) in proceedings instituted otherwise than under this section; or

(b) affects any restriction imposed by or under any other enactment whether public, local or private.

(6) In this section -

"controlled waters" has the same meaning as in Part II of this Act: and

"mine" has the same meaning as in the Mines and Quarries Act 1954.

APPENDIX TWO

PROPOSALS FOR EPA'90 SECTION 143 REGISTERS

PROPOSALS FOR EPA '90 SECTION 143 REGISTERS

The revised draft regulations which were issued by the DoE on 31 July 1992, made a variety of changes to original proposals, which were to be implemented in April 1992. The most acute change to the proposals is the reduced nature of the list of contaminative uses, estimated to cover 10 - 15% of the areas previously envisaged (DoE, 1992). Schedule 1 of the Draft Regulations lists the specified contaminative uses as follows:

1. Manufacture of gas, coke or bitumous material from coal.

2. Manufacture of refining of lead or steel or an alloy of lead or steel.

3. Manufacture of asbestos or asbestos products.

4. Manufacture, refining or recovery of petroleum or its derivatives, other than extraction from petroleum-bearing ground.

5. Manufacture, refining or recovery of other chemicals, excluding minerals.

6. Final deposit in or on land of household, commercial or industrial waste (within the meaning of Section 75 of the Act) other than waste consisting of ash, slag, clinker, rock, wood, gypsum, railway ballast, peat, bricks, tiles, concrete, other materials or dredging spoil; or where the waste is used as a fertilizer or in order to condition the land in some other beneficial manner.

7. Treatment at a fixed installation of household, commercial or industrial waste (within the meaning of section 75 of the Act) by chemical or thermal means.

8. Use of scrap metal store, within the meaning of Section 9(2) of the Scrap Metal Dealers Act 1964(a).

(Source DoE, July 1992)

APPENDIX THREE

CURRENT AND PROPOSED NRA R&D PROJECTS

CURRENT AND PROPOSED NRA R&D PROJECTS

POLLUTION RISK ASSESSMENT

Objectives: To utilise existing national databases on land use activities and catchment characteristics to produce GIS based pollution risk assessment tools which can be used to support policy and operational decisions.

CHEMICAL CONTAMINANTS

Objectives: To improve understanding of the chemistry and biochemistry of chemical contaminants plus their sources, transportation and fate, and to assess their significance for water quality management in order to facilitate their control and to develop response to existing problems of contamination.

POLLUTION POTENTIAL OF CONTAMINATED SITES (A REVIEW)

Objectives: To review procedures to assess groundwater pollution potential of contaminated sites and to relate these leaching tests to target levels for acceptability with respect to groundwater.

REMEDIAL TREATMENT METHODS FOR CONTAMINATED LAND

Objectives: To provide guidance to developers and regulatory bodies in the assessment, specification, supervision and achievement of effective and safe remediation of contaminated land using the most appropriate techniques.

METHODOLOGIES FOR THE PURIFICATION OF GROUNDWATERS CONTAMINATED BY SOLVENTS

Objectives: To assess the techniques and equipment available in the UK to clean-up groundwater contaminated by chlorinated solvents and to estimate the likely costs.

POLLUTION POTENTIAL OF CONTAMINATED SITES

Objectives: To assess the performance of a standardised leach test method when assessing water pollution potential in field conditions associated with contaminated land, thereby aiding judgement on the need and scale of remedial works required.

REFERENCES

1. Addison, R.F. (1989) Organochlorines and marine manual reproduction. Can, J., Fish Aquat Sci, Vol.46, pp 360-368.

2. British Standards Institute., (1989) BSI, Draft for Development 175. Code of practice for the identification of potentially contaminated land and its investigation

3. Bryce, A., (1992) Clean-Up Powers and Contamination: The New Regulatory Framework. In Contaminated Land: Key Environmental Issues for 1992 and Beyond. SBIM Environment Conference. February, 1992.

4. Department of the Environment., (1989) House of Commons Environment Committee Report "Contaminated Land" Minutes of Evidence, Vol I, p253 HMSO, 1990.

5. Department of the Environment., (1989) House of Commons Environment Committee Report "Contaminated Land" Minutes of Evidence, Vol I, p253 HMSO, 1990.

6. Department of the Environment., (1989) Derelict Land Survey 1998. HMSO, 1989.

7. Department of the Environment., (1992) River Quality: The Government's proposals. November 1992.

8. Environmental Resources Ltd., (1987) House of Commons Environment Committee Report "Contaminated Land". Minutes of Evidence, Vol II, p254. HMSO, 1990.

9. ENDS Report., (1991) "Contaminated Land Policy stutters forward" ENDS Report, 193, February 1991.

10. Harris, R.C., (1992) Consequences of Groundwater Pollution. In Groundwater Issues in Contaminated Land. EBS Conference, London, May,1992

11. Harris R.C. & Flavin, R.J., (1991) Contaminated Land: Implications for Water Pollution. Journal of IWEM, 5, p529-533, 1991.

12. Ironside Environmental Consultants., (1992) UK Strategy on Contaminated Land. Ironside Environmental Consultants, Edinburgh, 1992.

13	Kreager, D.,	(1991)	"Hazardous Waste Policies/Costs" Environment Report December 12, 1991, Washington D.C. 20004.
14	McLeod, G.,	(1991)	"Contaminated Land - The Regulatory Framework. Environmental Protection Bulletin 015, November 1991.
15	National Rivers Authority	(1992)	The Influence of Agriculture on Natural Waters in England and Wales. Water Quality Series, No. 6, NRA, 1992.
16	NSCA	(1992)	NSCA Pollution Handbook. NSCA, Brighton, 1992.
17	Pentreath, R.J.,	(1992)	Controlling the Inputs of Persistent Chemicals to Marine Waters from Land- Based Sources. British Assoc Science Festival. University of Southampton, August, 1992.
18	Reijnders, J. H.	(1986)	Reproductive failure in common seals feeding on fish farm polluted coastal waters. Nature, Vol 324, p456-457
19	University of Sheffield (Environmental Consultancy) and Richards, Moorehead and Laing.	(1994)	The reclamation and management of Metalliferous Mining Sites. London. HMSO (1994)